Biomaterials in Microencapsulation

Edited by Ashutosh Sharma

Published in London, United Kingdom

Biomaterials in Microencapsulation
http://dx.doi.org/10.5772/intechopen.1002081
Edited by Ashutosh Sharma

Contributors
Adina Magdalena Musuc, Arezou Pezhman, Charuvila T. Aravindakumar, Koulin Chen, Mariana Chelu, María Paulina Romero, Mengyan Nie, Myriam Alexandra González, O. K. Sasina Sai, Usha K. Aravind

First published in London, United Kingdom, 2024 by IntechOpen
IntechOpen is the global imprint of INTECHOPEN LIMITED, registered in England and Wales, registration number: 11086078, 167-169 Great Portland Street, London, W1W 5PF, United Kingdom

British Library Cataloguing-in-Publication Data
A catalogue record for this book is available from the British Library

Additional hard and PDF copies can be obtained from orders@intechopen.com

Biomaterials in Microencapsulation
Edited by Ashutosh Sharma
p. cm.
Print ISBN 978-0-85466-107-7
Online ISBN 978-0-85466-106-0
eBook (PDF) ISBN 978-0-85466-108-4

For EU product safety concerns:
IN TECH d.o.o., Prolaz Marije Krucifikse Kozulić 3, 51000 Rijeka, Croatia,
info@intechopen.com or visit our website at intechopen.com.

Meet the editor

Dr. Ashutosh Sharma is currently working as an associate professor in the Department of Physics, Amity Institute of Applied Sciences, Amity University Jharkhand, Ranchi, India. Previously, he worked as an assistant professor in the Department of Materials Science and Engineering, at Ajou University, Suwon, South Korea. He earned his Ph.D. in metallurgical and materials engineering from the Indian Institute of Technology (IIT) Kharagpur, a top-tier institution in India. His research interests include electrochemical deposition, lead-free soldering and brazing, additive manufacturing, high entropy alloys, gas sensors, and transparent nanoheaters. Dr. Sharma is a life member of various scientific and professional bodies. In a very short time, he has contributed more than 140 international journal articles, 14 patents, three book chapters, and one authored book so far. Recently, he was awarded the Extraction and Processing Division (EPD) Award by The Minerals, Metals and Materials (TMS, USA) in 2016 in recognition of his outstanding research in the area of nonferrous materials processing.

Contents

Preface

This book compiles cutting-edge research on biomaterials and encapsulating systems, which are critical in advancing a variety of medicinal and technological applications. The chapters delve further into new drug delivery mechanisms, encapsulation systems, and biomaterials' potential to transform sectors such as medicine, marine biology, and beyond. Encapsulation technology, in particular, provides novel approaches of delivering medicinal substances in a regulated and targeted manner, ensuring optimum efficacy while reducing unwanted effects.

The various chapters demonstrate the enormous potential of encapsulation technologies and biomaterials to solve a wide range of global concerns, from improving medication delivery and medical treatments to increasing the sustainability and efficiency of marine systems. This collection is an excellent resource for academics, practitioners, and students interested in the intersections of biotechnology, materials science, and applied treatments.

I would like to express our deepest gratitude to everyone at IntechOpen who helped bring this book to fruition, including the editors, contributors, coauthors, and publication staff. The novel experimental approach outlined in the book is provided with our sincere wish that it will pave the way for future innovations in materials science. Students at both the graduate and undergraduate levels, as well as researchers and engineers in a variety of fields, will find this book to be an invaluable resource. We wish for this book to guide the readers and hope it will help them develop novel biomaterials with desirable characteristics.

Dr. Ashutosh Sharma
Department of Physics,
Amity Institute of Applied Sciences,
Amity University Jharkhand,
Ranchi, Jharkhand, India

Chapter 1

Biomaterial in Microencapsulation: How Microencapsulation is Changing the Medicine World

Arezou Pezhman

Abstract

Stem cell therapy is one of the novel treatment. Cells possess self-renewal ability and the potential to differentiate into multiple lineages. Cell therapy has been studied in treatment of various diseases and injuries, such as cardiovascular diseases, brain disorders, musculoskeletal defects, osteoarthritis, and skin diseases. The application of cells can be a big challenge in treatment, and they die during transplants because of the unfavorable environments of injured or damaged tissues. A supportive environment can help cell survival, induce bio-activity, and enhance cell retention at the administered sites. Stem cell microencapsulation in biocompatible biomaterials can be a good supportive environment that lets cells grow properly. In this review, we discuss about new materials, their application for microencapsulation and how these materials can alter drug delivery and treatment of diseases. New natural and artificial materials optimize microencapsulation application and can be a novel solution for what scientist struggle with.

Keywords: biomaterial, stem cell, tissue engineering, drug delivery, hydrogel

1. Introduction

Stem cell therapy and biomaterial for drug delivery as novel therapies have recently offered new opportunities in clinical applications that permitted more precise and early diagnoses, less invasive and quicker procedures, and fewer and stay hospital visits. Stem cell therapy has been performed to treat various diseases and injuries, such as cardiovascular diseases, osteoarthritis, brain, neural, and musculoskeletal disorders [1–4]. Stem cells are capable of renewing themselves and have the potential to differentiate into multiple lineages, which include pluripotent stem cells, such as embryonic stem cells (ESCs), induced pluripotent stem cells (iPSCs), and multipotent stem cells such as hemopoietic stem cells (HSC), mesenchymal stem cells (MSCs), and adult stem cells (ASCs). Cell encapsulation is based on the immobilization of cells in favorable with synergistic intercellular interactions without being washed out or damaged due to the surrounding shear forces. Cell encapsulation technology can protect cells from hydrodynamic pressure and cell aggregation while allowing for the functional diffusion of nutrients, growth factors, and gases through the microcapsule matrix [5]. Ideal candidates for encapsulation would be non-immunogenic, non-tumorigenic, easy to access and form, well-specify, and reproducible. A lot of different cell types have been used for encapsulation, each with its unique advantages and limitations [6]. Several materials have been used in cell encapsulation including alginate [7], poly(lactic-*co*-glycolic acid)/poly(L-lactic acid) scaffolds [8], agarose [9], chitosan [10], poly ethylene glycol [11] and hyaluronic acid [12]. However, natural materials have some limitations, such as hard accessibility, finite source, rapid degradation, weak mechanical properties, and limited processability, which slow their clinical application. Also, the purity degree of some materials determines the biocompatibility of the microcapsules, so the purity degree and viability must be examined well [13]. Microencapsulation can offer a beneficial method of administering therapeutic materials and molecules that cannot be replaced by pharmacological substances, enabling novel solutions for the treatment of many diseases. Some applications of stem cell encapsulation in treatment of disease are listed in **Table 1**.

The long-term viability of microencapsulated cells, rate of polymer membrane degradation, pericapsular fibrosis formation tendency, and potential risk of immunogenicity to the polymers remain big challenges in microencapsulation. The encapsulating polymer must provide a barrier to prevent the immune system from recognizing the foreign cells [22]. A good material used for encapsulation should not interfere with the host cell's homeostasis. The direct side effect of a non-compatible material is type 1 hypersensitivity reaction, which is a non-specific body reaction that starts with the non-specific absorption of macrophages to the implant site [23]. The permeability of materials is one of the important features of encapsulation. The microcapsule membrane must have an appropriate pore size to supply essential nutrients necessary for cell survival, and it must also have enough immunoprotection to forbid contact with host immune cells [6]. Mechanical resistance of a material is critical to ensure cell immunoprotection and molecule release time period define the deformation and rupture of the capsules under external load [24]. There are some methods and techniques for microencapsulation application, as highlighted in **Table 2** [25].

Chemical methods of microencapsulation include solvent evaporation, interfacial crosslinking, interfacial polycondensation, *in situ* polymerization, and matrix polymerization which usually undergo heat reduction after mixing and heating [26]. Physical methods include spray drying, fluid-bed/pan coating, centrifugal extrusion, vibrating nozzle, and spinning disk microencapsulation. The dispersion of an oil core material or water-soluble active ingredient into a concentrated coating material is used to prepare for spray drying [27]. In fluid-bed coating, solid particles are mixed with a dry coating material that is heated to surround the particle cores [28].

Bone [14]
CNS [15]
Cartilage [16]
Cardiovascular [17]
Cancer [18]
Diabetes [19]
Liver [20]
Skin [21]

Table 1.
Application of stem cell.

Mechanical (physical) methods
Solvent evaporation
Spray drying
Droplet freezing
Airflow or fluidized bed
Droplet gelation
Centrifugation
Extrusion
Chemical methods
In situ and interfacial polycondensation
Polymerization
Gelation
Physicochemical methods
Methods using supercritical fluid
Methods using emulsification
Coacervation
Thermal gelation

Table 2.
Methods for microencapsulation.

The microcapsule core and coating materials are both immiscible with each other, so centrifugal extrusion is pressed into the concentric nozzles, forming a flow that is crushed into droplets following the clearing of the nozzle [25]. The liquid material to be encapsulated is extruded through a nozzle at a specific flow rate in a vibrating nozzle [29]. In a spinning disk, a mixture is configured with the liquid microcapsule coating material and the material for the internal core of the microcapsule [25]. Physicochemical encapsulation techniques involve supercritical fluid technology, coacervation, polyelectrolyte complexation, and ionotropic gelation. Ionotropic gelation works through the polyelectrolyte's ability to attach when in the presence of counterions, resulting in their gelation [30]. In polyelectrolyte complexation, adding polycations or polyelectrolytes can be used to further improve the mechanical strength and permeability of the gelated beads [25]. Coacervation technique that relies on polymer-polymer incompatibility, addition of a salt or alcohol into the polymeric mixture or modification of the aqueous phase pH, lead to crosslinking of polymers [31]. Supercritical fluid technology is relied on the formation of particles that are monodispersed with the capability to form nanosized particles [32]. A lot of synthetic and natural materials are being used in microencapsulation. Some of these materials and their application are described below.

2. Alginate

Alginates are natural polysaccharides, polyanionic polymers, and unbranched copolymers of D-mannuronic acid and L-guluronic acid linked by β (1-4) glycosidic bonds [33–35]. Alginate is derived from various species of brown algae cell walls such as hyperborea, Ecklonia maxima, Ascophyllum nodosum, Eisenia bicyclis, Laminaria, Macrocystis pyrifera, etc., and from various species of bacteria: Azotobacter and Pseudomonas [36, 37]. Alginate is a biodegradable and biocompatible polymer with low immunogenicity in human body. Alginate's ability to form hydrogel makes it a useful material for the delivery of medicines and cellular immobilization [38, 39]. Because of gel and hydrogel-forming of alginate, it can be easily synthesized and manipulated and can control the speed of release of active compounds from pharmaceutical systems [40, 41]. Chemically modified alginic acid, because of its strong polyanions feature, was found to inhibit fibrotic formation after transplantation by inactivating an immune response, while cationic materials tend to induce inflammatory reaction [42]. Alginate helps in treatment of many diseases. Some of the alginate application include intestinal inflammation and colitis in inflammatory bowel disease, diabetes, hemophilia, cancer, renal failure, liver, and musculoskeletal diseases [18, 43–47]. Also, because of gel-forming properties of alginate, simple structure, simple raw materials, low toxicity, and mild processing, alginate can be a practical material in drug delivery such as probiotic and anticancer drugs [48, 49].

3. Agarose

Agarose, usually derived from red algae and seaweed, includes a galactose-based structure. Its ability to undergo reversible gelation in response to temperature makes it a functional biomaterial for tissue engineering applications [50]. Agarose is thermally responsive [50] and very similar to the extracellular matrix of natural cell, making it suitable for cell encapsulation. Agarose has been used in medical applications widely because of its functional features including controlled self-gelling properties, water solubility, adjustable mechanical properties, non-immunogenic nature and it is biocompatibility [51]. Agarose, based on its stiffness and functional groups, can support cellular adhesion, proliferation, and activity. In tissue engineering, agarose hydrogels were combined with cells and applied for the regeneration of several structures such as cartilage [52], bone [53], tendons [54], neural system [55], cardiac regeneration, wound healing [56] and cornea. The limitation of using agarose is that mammalian cells cannot degrade agarose, so it can affect gene expression and cell phenotype [57].

4. Chitosan

Chitosan is a derivative of the shell waste of crab, shrimp, and crawfish and a polysaccharide composed of (1, 4)-linked 2-amino-deoxy-b-D-glucan [58, 59]. Chitosan coating a unique cationic character, high biocompatibility, non-toxicity, and biodegradability that make it a functional material in the pharmaceutical industry and tissue engineering, improving improved mechanical strength and a strong barrier function [60]. Furthermore, chitosan properties, such as low solubility in water and organic solvents, can be improved by chemical modification. Also, chitosan has great potential to be used in common applications for the preparation of microspheres, nanospheres, and microcapsules. Chitosan is a cationic biopolymer and has intrinsic antimicrobial activity, besides other potentialities like antioxidant, antitumor, and immune modulator, with an effective drug delivery agent. Medical application of chitosan improves disease treatment, especially in cartilage, bone, and nerve repair, and upgrades tissue engineering methods in their treatment [61–63]. Wound healing, Gene Silencing in Disease Vector Mosquito Larvae, obesity, and cardiovascular disease treatment are other uses of chitosan and derivatives [64–67]. Chitosan is widely used for drug encapsulation, growth factors, antimicrobials, painkillers, and antitumoral or anti-inflammatory drugs [68–72]. Also, chitosan and chitosan derivatives show antimicrobial activity against different microorganisms, including bacteria, filamentous fungi, and yeast, that can be a solution in bacterial resistance to antibiotics [73]. One of the important industrial applications of chitosan is its role in water filtering, which helps in the adsorption of pollutants in water [74].

5. Dextran

One of the suitable materials for use as a scaffold in tissue engineering is dextran [75]. Dextran is a neutral, biodegradable polysaccharide formed from sucrose by bacteria. Anti-thrombolytic and bioadhesive properties make it one of the best scaffolds in tissue engineering [76]. Also, a specific characteristic feature of dextran is The active hydroxyl groups of dextran can be chemically modified to incorporate various functional groups, so dextran has been chemically engineered to form various scaffolds, including spheres [77–79], tubules [80], and hydrogel [81, 82]. Dextran has been used a lot in biomedical and pharmaceutical applications like decreasing vascular thrombosis [83, 84], reducing inflammatory response [85], preventing ischemia-reperfusion injury in organ transplantation [86–88], promoting wound healing [89, 90], and enhancing chondrogenesis and cartilage regeneration [91]. Due to its excellent physicochemical properties and biocompatibility, dextran has been assumed one of the best materials for drug delivery and transport of therapeutic agents [92, 93].

6. Gellan gum

Gellan gum is one of the best polymers in tissue engineering and drug delivery because of its biodegradability, biocompatibility, and sustainability features [94, 95]. Gellan gum is formulated by bacterial exopolysaccharide fermentation from *Pseudomonas elodea* or *Sphingomonas elodea*, negatively charged, and contains linear exopolysaccharides consisting of four repeating carbohydrates in the main chain [96, 97]. Its ability to undergo sol-gel transition and form a solid gel upon contact with cations can be used to prepare drug-loaded matrices [98]. This features make gellan gum a perfect material in tissue engineering and regenerative medicine, such as its wide utilization in cartilage [99–101], bone [102–104], osteochondral disorders [105], spinal cord injury [106], wound healing [107], diabetic ulcer [108] and retinal injury [109]. However, native gellan gum has some limitations, requiring high temperature (60–90°C) for water dissolution and high concentration for the production of stable physical hydrogel that limits production procedure and suitability for direct cell encapsulation [110].

7. Matrigel

Matrigel, an extract derived from the EHS (Engelbreth-Holm-Swarm) tumor, contains all of the known major components of many tissue basement membranes that was found by Dr. Engelbreth-Holm from spontaneous mouse tumor. Matrigel contain all of the components present in basement and is very biologically active [111]. The components of Matrigel are four major basement membrane extracellular matrix (ECM) proteins: laminin, collagen IV, entactin, and the heparin sulfate proteoglycan perlecan [112] and also contain collagen I, XVIII, VI [113] and III and tumor-derived proteins, including growth factors, such as fibroblast growth factors (FGFs) and transforming growth factor (TGF) family peptides (for example, TGF-β) [114, 115] and enzymes, such as matrix metalloproteinases (MMPs) [111, 116]. Matrigel undergoes gelation at temperatures in the range of 22–37°C and is used to culture and cell differentiation such as human pluripotent stem cells (hPSCs) [117], neural tissue [118, 119], and cardiomyocytes [120]. Matrigel also has been used for endothelial tubulogenesis [121, 122], an organoid assembly such as inner ear organoid [123], and applied for promoting angiogenesis [124], cell differentiation in tissue engineering [125], and skin repair [126]. Matrigel application is limited because of the presence of xenogenic contaminants [114, 127], so it needs to be used with caution.

8. Gelatin methacrylate

Gelatin methacrylate (GelMA) was developed by modifying the reactive side groups of gelatin using glycidyl methacrylate [128] that allowed photo-crosslinking so as to provide customizable mechanical properties [129]. GelMA can be manufactured through several methods, such as soft lithography [130], electrospray [131], three-dimensional printing [132], microfluidic chip [133], and emulsion [134]. However, UV light exposure during crosslinking can reduce the survival rate of cells and damage organisms [135–137]. GelMA has a dynamics macromolecular network and biomimic microenvironment that has a high water content [138, 139]. Hydrogels like GelMA are planned to be similar to the characteristics of extracellular matrix (ECM), and three-dimensional (3D) structure supports for cellular growth and tissue formation [140] and also used for cell-cell interactions, proliferation, migration [141], and controlled differentiation [142] in tissue engineering. Biocompatibility, biodegradability and low cost makes GelMA [143] useful in drug delivery [144], cell delivery [145], gene delivery [146], vaccine delivery [147], wound healing [148, 149], and tissue regeneration [150]. Also, GelMA has broad applications in biomedical field such as corneal regeneration [151], bone repair [152, 153], myocardial infarction [154, 155], hemostasis control [156], and vascular regeneration [157, 158].

9. Polyethylene glycol

Polyethylene glycol (PEG) is a product that is used in both industrial and pharmaceutical industries. PEG is a low-toxic, hydrophobic, semicrystalline polyether that is used in drug delivery systems and tissue engineering scaffold formation [159]. Biocompatibility and hydrophilic feature with properties that limit antigenicity, immunogenicity, cell adhesion, and protein binding [160] makes PEG a successful scaffolding material for 3D printing in tissue engineering [161]. The essential molecular structure of PEG is PEG diol surrounded with two hydroxyl end groups, which can be switched into other functional groups such as carboxyl, amine, vinyl sulfane, methoxyl, thiol, azide, acetylene, and acrylate [162, 163]. Photopolymerization is the most common technique to make PEG hydrogel [164]. The photoinduced crosslinking strategy enhances the crosslinking parameters and biocompatibility by functionalizing them with acrylic or methacrylic groups [165]. Photopolymerization can mix synthetic and modified polymers like polyethylene glycol diacrylate (PEGDA) with natural polymers to form a double network hydrogel for better results [166]. Also, synthetic polymers, such as modified PEG, are able to mimic the biochemical properties of proteins [167, 168]. Because of these features, PEG has become impressing material for bone repair, nerve tissue regeneration [169, 170], and bioprinting applications [171].

10. Polycaprolactone

PCL is a semicrystalline aliphatic polyester that is biodegradable, biocompatible, and non-toxic. PCL exhibits excellent chemical and solvent resistance, good toughness, low glass transition (−60°C), and melting temperature (60°C) [172]. The distribution of water molecules into formless regions causing hydrolytic breakdown of ester bonds can degrade PCL, initially in the formless part, followed by the crystalline domain [173]. PCL has gained lot of importance in the design of biomaterials/green materials in areas like wound dressing, contraceptives, and dentistry, along with nonmedical fields like food, packaging, and environment [174, 175]. Because of PCL's high permeability to many drugs, excellent biocompatibility, slow biodegradability, and bioresorbability, PCL is used for long-term drug delivery such as anticancer [176, 177], antipsychotic [178], non-steroidal anti-inflammatory [179, 180], anti-hypertensive drugs [181] and, others. PCL is also one of the most common materials in tissue engineering and 3D scaffolds. Polycaprolactone features such as mechanical properties with high flexibility and great elongation make it an excellent material for the preparation of scaffolds for bone repair [182] and wound dressing [183]. 3D-printed medical devices are an advanced application of PCL. PCL airway devices [184], congenital heart defects, gastric wall damage (hollow organ), and periodontal repair are affecting millions of people around the world [185–187].

11. Polyvinylpyrrolidone

Polyvinylpyrrolidone (PVP) is a hygroscopic, amorphous, synthetic polymer consisting of linear 1-vinyl-2-pyrrolidinone groups [188]. Polyvinylpyrrolidone is a biodegradable, water-soluble polymer that is biocompatible and non-toxic. It also has a stabilizing effect on suspensions and emulsions [189, 190]. PVP is widely used in the food industry, cosmetics, pharmaceutical, and biomedical applications [191, 192] as well as drug delivery such as gene delivery, oral, topical, transdermal, and ocular administration [193–195]. PVP constitutes a group of crosslinked 3D networks of hydrophilic polymer chains, which are capable of holding large amounts of water without dissolving [196]. PVP structure is similar to living tissues and tissue matrix so that it can be used for the fabrication of tissue engineering coatings and wound dressings, bone and cartilage repairing systems, drug delivery systems, and other body-contacting devices products, such as contact lenses [196–198]. However, there have been documented cases of allergic reactions to PVP, particularly regarding subcutaneous use and situations where the PVP has come in contact with autologous serum and mucous membranes [199–201]. Allergic reactions of PVP commonly happen because of conjunction with other chemicals, such as iodine [202].

12. Poly(l-lactic) acid

Polylactic acid (PLA) is a biodegradable material derived from lactic acid (LA) and produced from reproducible origine such as corn, wheat, and straw [203, 204]. PLA is classified as an aliphatic polyester because of the ester bonds that connect the monomer units. It is made by the bacterial fermentation of carbohydrates using optimized strains of the genus Lactobacillus [205]. PLA is not altered by solvent bulging and dissolution during industrial fabrication. Processing temperature is generally 170–230°C, so it is also a proper material for spinning, biaxial stretching, extrusion, and injection blow-molding processing methods [206–208]. PLA is used as bio-based plastic and reduces plastic-induced environmental problems [209]. PLA has been widely used as a scaffold in tissue engineering. PLA-based scaffold can be manufactured by controlling the related parameters such as crystallinity, molecular weight, copolymer ratio, essential viscosity, and residual monomers to reach the appropriate physical properties so can stimulate isolated cells to regenerate tissues and act as a drug delivery carrier [210, 211]. Also, PLA has got an important role in biomedical applications such as suture threads, bone fixation screws, and devices for drug delivery, just to scratch the surface [212]. In addition to the musculoskeletal tissue engineering application of PLA, it has been used in nervous, cardiovascular, and cutaneous application [213–215]. The usage of PLA, especially in the biomedical field, is still limited by its low biodegradability and hydrophobicity [216].

13. Poly(lactic-*co*-glycolic) acid

Poly(lactic-*co*-glycolic) acid (PLGA) is a copolymer with lactic and glycolic acid repeat units with linear structure, which can be formed as a block-*co*-polymer or statistical polymer. The body can degrade both monomers into carbon dioxide and water via the Krebs cycle, which could be achieved from the fermentation of corn or other grains, and the glycolic acid part could be created from the biochemical enzymatic reaction or by the chemical combination of chloroacetic acid and sodium hydroxide [217–219]. Polymer characteristic features, such as degradation kinetics, rheological properties, thermal properties, mechanical properties, water uptake, and release profiles, can be determined by modification of the LA/GA ratio, terminal group, and constructed parameters of PLGA that facilitates the engineering of this polymer. Higher PGA content leads to quicker metabolization rates, with the exception of an equal ratio of PLA/PGA, which shows the fastest degradation rate, while higher PGA substances lead to increased degradation [220]. PLGA metabolizes by hydrolysis of its ester linkages [221]. PLGA can be dissolved by a wide range of popular solvents, such as tetrahydrofuran, chlorinated solvents, acetone or ethyl acetate, though pure polyglycolic acid and polylactic acid show poor solvability [222]. PLGA is an attractive provider of devices for multidrug distribution and multiplication operations. Numerous studies have also reported successful applications in antibiotics, antiseptics, anticancer, and imaging medications [223, 224]. PLGA can be easily processed and fabricated in various forms and sizes as nanoparticles (microspheres, microcapsules, nanocapsules, and nanospheres) [225]. One of the most generally used biodegradable synthetic polymers for 3D scaffolds in tissue engineering is PLGA. PLGA grafts can be embedded with many regenerative factors to boost healing, such as growth factors, stem cells, and drugs, that make them suitable for bone repair and cartilage regeneration [226–228].

DOI: http://dx.doi.org/10.5772/intechopen.1005202

14. Conclusion

The word microcapsules implies the membrane-surrounded particles or droplets distributed in a solid matrix. Microencapsulation has gained importance in the fields of cell and tissue engineering. The encapsulation system provides free exchange of what cells need to grow such as nutrients and oxygen between the loaded cells and their surrounding environment while preventing the removal and elimination of the loaded cells. The choice of an encapsulating agent depends on several factors, such as biocompatibility, processing, physicochemical properties of the substance to be encapsulated, and degradation prevention. Microencapsulation is a functional technology for the immunoprotection of biological material to avoid host immune response and protect cells from immune suppression. Polymeric biomaterial, either of natural or synthetic origin, must have high biocompatibility and biotolerability, similarity to the extracellular matrix (ECM) of human tissues and their composition, that simulate the biological environments of the human body. Also, microencapsulation has become increasingly important in the fields of pharmaceutical engineering and improved drug delivery systems. Microencapsulation technology has developed industrial materials such as food industry. In fact, microencapsulation can be a good solution to face the problem that we struggle with. This technology can help to cure many diseases, upgrade treatment, and reduce complications and treatment costs. Microencapsulation can provide better drug delivery with lower complications and side effects. Also, a new method of microencapsulation can promote potential and profitability in the industry. Choosing the right material and its properties is important to get maximal utilization. Furthermore, the combination of materials increases its potential. Two main groups of microencapsulation materials are natural and synthetic. Natural materials have high biocompatibility, degradability, and similarity to human extracellular matrix (ECM) but have limited sources and harder accessibility. In spite of natural material privileges, synthetic materials are preferred because of easy manufacturing, unlimited accessibility, and good results in medical applications. The best result can be reached by choosing the appropriate material.

15. Future perspective

Nowadays, in spite of all developments in the medical field, humans suffer from disease side effects. Microencapsulation can be a novel way to overcome disease treatment challenges. Microencapsulation with or without biocontent can accelerate healing and minimize side effects. However, microencapsulation is a novel technique, and there is a long way to go to find its place as a routine medical application. By exploring and inventing new material, new instruction, and application methods, we can accelerate microencapsulation usage in all medical fields.

Author details

Arezou Pezhman
Department of Internal Medicine, Zahedan Azad University of Medical Science, Zahedan, Iran

*Address all correspondence to: arezou_pejman@yahoo.com

References

[1] Segers VF, Lee RT. Stem-cell therapy for cardiac disease. Nature. 2008;**451**(7181):937-942

[2] Kuo TK, Hung SP, Chuang CH, Chen CT, Shih YRV, Fang SCY, et al. Stem cell therapy for liver disease: Parameters governing the success of using bone marrow mesenchymal stem cells. Gastroenterology. 2008;**134**(7):2111-2121. e3

[3] Crevensten G, Walsh AJ, Ananthakrishnan D, Page P, Wahba GM, Lotz JC, et al. Intervertebral disc cell therapy for regeneration: Mesenchymal stem cell implantation in rat intervertebral discs. Annals of Biomedical Engineering. 2004;**32**:430-434

[4] Davatchi F, Abdollahi BS, Mohyeddin M, Shahram F, Nikbin B. Mesenchymal stem cell therapy for knee osteoarthritis. Preliminary report of four patients. International Journal of Rheumatic Diseases. 2011;**14**(2):211-215

[5] Zimmermann H, Wählisch F, Baier C, Westhoff M, Reuss R, Zimmermann D, et al. Physical and biological properties of barium cross-linked alginate membranes. Biomaterials. 2007;**28**(7):1327-1345

[6] Uludag H, De Vos P, Tresco PA. Technology of mammalian cell encapsulation. Advanced Drug Delivery Reviews. 2000;**42**(1-2):29-64

[7] Siti-Ismail N, Bishop AE, Polak JM, Mantalaris A. The benefit of human embryonic stem cell encapsulation for prolonged feeder-free maintenance. Biomaterials. 2008;**29**(29):3946-3952

[8] Levenberg S, Huang NF, Lavik E, Rogers AB, Itskovitz-Eldor J, Langer R. Differentiation of human embryonic stem cells on three-dimensional polymer scaffolds. Proceedings of the National Academy of Sciences. 2003;**100**(22):12741-12746

[9] Dang S. Scalable bioprocess for controlled differentiation of embryonic stem cells to hematopoietic progenitors. 2004

[10] Li Z, Leung M, Hopper R, Ellenbogen R, Zhang M. Feeder-free self-renewal of human embryonic stem cells in 3D porous natural polymer scaffolds. Biomaterials. 2010;**31**(3):404-412

[11] Siltanen C, Diakatou M, Lowen J, Haque A, Rahimian A, Stybayeva G, et al. One step fabrication of hydrogel microcapsules with hollow core for assembly and cultivation of hepatocyte spheroids. Acta Biomaterialia. 2017;**50**:428-436

[12] Gerecht S, Burdick JA, Ferreira LS, Townsend SA, Langer R, Vunjak-Novakovic G. Hyaluronic acid hydrogel for controlled self-renewal and differentiation of human embryonic stem cells. Proceedings of the National Academy of Sciences. 2007;**104**(27):11298-11303

[13] Orive G, Ponce S, Hernández RM, Gascón AR, Igartua M, Pedraz JL. Biocompatibility of microcapsules for cell immobilization elaborated with different type of alginates. Biomaterials. 2002;**23**(18):3825-3831

[14] Yang L, Liu Y, Sun L, Zhao C, Chen G, Zhao Y. Biomass microcapsules with stem cell encapsulation for bone repair. Nano-micro Letters. 2021;**14**(1):4

[15] Lopez-Mendez TB, Santos-Vizcaino E, Pedraz JL, Hernandez RM, Orive G. Cell microencapsulation technologies for sustained drug delivery: Clinical trials and companies. Drug Discovery Today. 2021;**26**(3):852-861

[16] Liu Y, Wang M, Luo Y, Liang Q, Yu Y, Chen F, et al. Enhancing stem cell therapy for cartilage repair in osteoarthritis-A hydrogel focused approach. Gels (Basel, Switzerland). 2021;**7**(4):263

[17] Karpov AA, Puzanov MV, Ivkin DY, Krasnova MV, Anikin NA, Docshin PM, et al. Non-inferiority of microencapsulated mesenchymal stem cells to free cells in cardiac repair after myocardial infarction: A rationale for using paracrine factor(s) instead of cells. International Journal of Experimental Pathology. 2019;**100**(2):102-113

[18] Shah K. Encapsulated stem cells for cancer therapy. Biomatter. 2013;**3**(1):1-7

[19] Neumann M, Arnould T, Su B-L. Encapsulation of stem-cell derived β-cells: A promising approach for the treatment for type 1 diabetes mellitus. Journal of Colloid and Interface Science. 2023;**636**:90-102

[20] Liu P, Mao Y, Xie Y, Wei J, Yao J. Stem cells for treatment of liver fibrosis/cirrhosis: Clinical progress and therapeutic potential. Stem Cell Research & Therapy. 2022;**13**(1):356

[21] Moyer HR, Kinney RC, Singh KA, Williams JK, Schwartz Z, Boyan BD. Alginate microencapsulation technology for the percutaneous delivery of adipose-derived stem cells. Annals of Plastic Surgery. 2010;**65**(5):497-503

[22] Marikar SN, El-Osta A, Johnston A, Such G, Al-Hasani K. Microencapsulation-based cell therapies. Cellular and Molecular Life Sciences. 2022;**79**(7):351

[23] Anderson JM. Inflammatory response to implants. ASAIO Transactions. 1988;**34**(2):101-107

[24] Sanz Nogués C, Thomas D, Pandit A, O'Brien T. Stem Cell Microencapsulation for Therapeutic Angiogenesis. United States: CRC Press; 2013. pp. 386-424

[25] Tomaro-Duchesneau C, Saha S, Malhotra M, Kahouli I, Prakash S. Microencapsulation for the therapeutic delivery of drugs, live mammalian and bacterial cells, and other biopharmaceutics: Current status and future directions. Journal of Pharmaceutics. 2013;**2013**:103527

[26] Li M, Rouaud O, Poncelet D. Microencapsulation by solvent evaporation: State of the art for process engineering approaches. International Journal of Pharmaceutics. 2008;**363**(1-2):26-39

[27] Mok H, Park TG. Water-free microencapsulation of proteins within PLGA microparticles by spray drying using PEG-assisted protein solubilization technique in organic solvent. European Journal of Pharmaceutics and Biopharmaceutics. 2008;**70**(1):137-144

[28] Singh M, Hemant K, Ram M, Shivakumar H. Microencapsulation: A promising technique for controlled drug delivery. Research in Pharmaceutical Sciences. 2010;**5**(2):65

[29] Haas PA. Formation of uniform liquid drops by application of vibration to laminar jets. Industrial & Engineering Chemistry Research. 1992;**31**(3):959-967

[30] Patil J, Kamalapur M, Marapur S, Kadam D. Ionotropic gelation and

polyelectrolyte complexation: The novel techniques to design hydrogel particulate sustained, modulated drug delivery system: A review. Digest Journal of Nanomaterials and Biostructures. 2010;**5**(1):241-248

[31] Mohanty B, Bohidar H. Systematic of alcohol-induced simple coacervation in aqueous gelatin solutions. Biomacromolecules. 2003;**4**(4):1080-1086

[32] Yeo S-D, Kiran E. Formation of polymer particles with supercritical fluids: A review. The Journal of Supercritical Fluids. 2005;**34**(3):287-308

[33] Tønnesen HH, Karlsen J. Alginate in drug delivery systems. Drug Development and Industrial Pharmacy. 2002;**28**(6):621-630

[34] Van Vlierberghe S, Graulus G-J, Samal SK, Van Nieuwenhove I, Dubruel P. Porous hydrogel biomedical foam scaffolds for tissue repair. In: Biomedical Foams for Tissue Engineering Applications. Sawston, Cambridge: Elsevier; 2014. pp. 335-390

[35] Asgari S, Pourjavadi A, Licht TR, Boisen A, Ajalloueian F. Polymeric carriers for enhanced delivery of probiotics. Advanced Drug Delivery Reviews. 2020;**161**:1-21

[36] Phillips GO, Williams PA. Handbook of Hydrocolloids. Amsterdam, Netherlands: Elsevier; 2009

[37] Sachan NK, Pushkar S, Jha A, Bhattcharya A. Sodium alginate: The wonder polymer for controlled drug delivery. Journal of Pharmacy Research. 2009;**2**(8):1191-1199

[38] Sellimi S, Younes I, Ayed HB, Maalej H, Montero V, Rinaudo M, et al. Structural, physicochemical and antioxidant properties of sodium alginate isolated from a Tunisian brown seaweed. International Journal of Biological Macromolecules. 2015;**72**:1358-1367

[39] Agüero L, Zaldivar-Silva D, Peña L, Dias ML. Alginate microparticles as oral colon drug delivery device: A review. Carbohydrate Polymers. 2017;**168**:32-43

[40] Ning H, Lu L, Xu J, Lu L, Pan L, Lin Z. Development of sodium alginate-based antioxidant and antibacterial bioactive films added with IRMOF-3/carvacrol. Carbohydrate Polymers. 2022;**292**:119682

[41] Pawar SN, Edgar KJ. Alginate derivatization: A review of chemistry, properties and applications. Biomaterials. 2012;**33**(11):3279-3305

[42] Bhatia SR, Khattak SF, Roberts SC. Polyelectrolytes for cell encapsulation. Current Opinion in Colloid & Interface Science. 2005;**10**(1-2):45-51

[43] Pi Y, Zhang X, Wu Y, Wang Z, Bai Y, Liu X, et al. Alginate alleviates dextran sulfate sodium-induced colitis by promoting bifidobacterium animalis and intestinal hyodeoxycholic acid synthesis in mice. Microbiology Spectrum. 2022;**10**(6):e02979-e02922

[44] Basta G, Montanucci P, Calafiore R. Microencapsulation of cells and molecular therapy of type 1 diabetes mellitus: The actual state and future perspectives between promise and progress. Journal of Diabetes Investigation. 2021;**12**(3):301-309

[45] Chen Z, Hu Q, Gu Z. Leveraging engineering of cells for drug delivery. Accounts of Chemical Research. 2018;**51**(3):668-677

[46] Doré E, Legallais C. A new concept of bioartificial liver based on

a fluidized bed bioreactor. Therapeutic Apheresis: Official Journal of the International Society for Apheresis and the Japanese Society for Apheresis. 1999;**3**(3):264-267

[47] Khatab S, Leijs MJ, van Buul G, Haeck J, Kops N, Nieboer M, et al. MSC encapsulation in alginate microcapsules prolongs survival after intra-articular injection, a longitudinal in vivo cell and bead integrity tracking study. Cell Biology and Toxicology. 2020;**36**(6):553-570

[48] Wang X, Gao S, Yun S, Zhang M, Peng L, Li Y, et al. Microencapsulating alginate-based polymers for probiotics delivery systems and their application. Pharmaceuticals (Basel, Switzerland). 2022;**15**(5):644

[49] He L, Shang Z, Liu H, Yuan ZX. Alginate-based platforms for cancer-targeted drug delivery. BioMed Research International. 2020;**2020**:1487259

[50] Zarrintaj P, Manouchehri S, Ahmadi Z, Saeb MR, Urbanska AM, Kaplan DL, et al. Agarose-based biomaterials for tissue engineering. Carbohydrate Polymers. 2018;**187**:66-84

[51] Irastorza-Lorenzo A, Sánchez-Porras D, Ortiz-Arrabal O, de Frutos MJ, Esteban E, Fernández J, et al. Evaluation of marine agarose biomaterials for tissue engineering applications. International Journal of Molecular Sciences. 2021;**22**(4):1923

[52] Salati MA, Khazai J, Tahmuri AM, Samadi A, Taghizadeh A, Taghizadeh M, et al. Agarose-based biomaterials: Opportunities and challenges in cartilage tissue engineering. Polymers. 2020;**12**(5):1150

[53] Hasan ML, Padalhin AR, Kim B, Lee BT. Preparation and evaluation of BCP-CSD-agarose composite microsphere for bone tissue engineering. Journal of Biomedical Materials Research Part B: Applied Biomaterials. 2019;**107**(7):2263-2272

[54] González-Quevedo D, Díaz-Ramos M, Chato-Astrain J, Sánchez-Porras D, Tamimi I, Campos A, et al. Improving the regenerative microenvironment during tendon healing by using nanostructured fibrin/agarose-based hydrogels in a rat Achilles tendon injury model. The Bone & Joint Journal. 2020;**102-b**(8):1095-1106

[55] Stokols S, Sakamoto J, Breckon C, Holt T, Weiss J, Tuszynski MH. Templated agarose scaffolds support linear axonal regeneration. Tissue Engineering. 2006;**12**(10):2777-2787

[56] Miguel SP, Ribeiro MP, Brancal H, Coutinho P, Correia IJ. Thermoresponsive chitosan–agarose hydrogel for skin regeneration. Carbohydrate Polymers. 2014;**111**:366-373

[57] Bissell MJ, Barcellos-Hoff MH. The influence of extracellular matrix on gene expression: Is structure the message? Journal of Cell Science Supplement. 1987;**8**:327-343

[58] Anal AK, Singh H. Recent advances in microencapsulation of probiotics for industrial applications and targeted delivery. Trends in Food Science & Technology. 2007;**18**(5):240-251

[59] Aider M. Chitosan application for active bio-based films production and potential in the food industry. LWT-Food Science and Technology. 2010;**43**(6):837-842

[60] Heidebach T, Först P, Kulozik U. Microencapsulation of probiotic cells for food applications. Critical Reviews in Food Science and Nutrition. 2012;**52**(4):291-311

[61] Chameettachal S, Murab S, Vaid R, Midha S, Ghosh S. Effect of visco-elastic silk–chitosan microcomposite scaffolds on matrix deposition and biomechanical functionality for cartilage tissue engineering. Journal of Tissue Engineering and Regenerative Medicine. 2017;**11**(4):1212-1229

[62] Li Q, Zhou G, Yu X, Wang T, Xi Y, Tang Z. Porous deproteinized bovine bone scaffold with three-dimensional localized drug delivery system using chitosan microspheres. Biomedical Engineering Online. 2015;**14**(1):1-15

[63] Jian R, Yixu Y, Sheyu L, Jianhong S, Yaohua Y, Xing S, et al. Repair of spinal cord injury by chitosan scaffold with glioma ECM and SB 216763 implantation in adult rats. Journal of Biomedical Materials Research Part A. 2015;**103**(10):3259-3272

[64] Rhoades J, Roller S. Antimicrobial actions of degraded and native chitosan against spoilage organisms in laboratory media and foods. Applied and Environmental Microbiology. 2000;**66**(1):80-86

[65] Zhang X, Mysore K, Flannery E, Michel K, Severson DW, Zhu KY, et al. Chitosan/interfering RNA nanoparticle mediated gene silencing in disease vector mosquito larvae. JoVE (Journal of Visualized Experiments). 2015;**97**:e52523

[66] Heber D. Herbal preparations for obesity: Are they useful? Primary Care: Clinics in Office Practice. 2003;**30**(2):441-463

[67] Yu Y, Luo T, Liu S, Song G, Han J, Wang Y, et al. Chitosan oligosaccharides attenuate atherosclerosis and decrease non-HDL in ApoE−/−mice. Journal of Atherosclerosis and Thrombosis. 2015;**22**(9):926-941

[68] Zeng W, Chang Z, Liu Z, Zhu L, Wang M, Hao D, et al. Controlled delivery of bioactive BDNF for potential treatment of peripheral nerve injury. Polymer Degradation and Stability. 2020;**181**:109296

[69] Khare T, Mahalunkar S, Shriram V, Gosavi S, Kumar V. Embelin-loaded chitosan gold nanoparticles interact synergistically with ciprofloxacin by inhibiting efflux pumps in multidrug-resistant *Pseudomonas aeruginosa* and *Escherichia coli*. Environmental Research. 2021;**199**:111321

[70] Basu SK, Kavitha K, Rupeshkumar M. Evaluation of ionotropic cross-linked chitosan/gelatin B microspheres of tramadol hydrochloride. AAPS PharmSciTech. 2011;**12**:28-34

[71] Dawoud M. Chitosan coated solid lipid nanoparticles as promising carriers for docetaxel. Journal of Drug Delivery Science and Technology. 2021;**62**:102409

[72] Öztürk AA, Kıyan HT. Treatment of oxidative stress-induced pain and inflammation with dexketoprofen trometamol loaded different molecular weight chitosan nanoparticles: Formulation, characterization and anti-inflammatory activity by using in vivo HET-CAM assay. Microvascular Research. 2020;**128**:103961

[73] Raafat D, Sahl HG. Chitosan and its antimicrobial potential–a critical literature survey. Microbial Biotechnology. 2009;**2**(2):186-201

[74] Vakili M, Rafatullah M, Salamatinia B, Abdullah AZ, Ibrahim MH, Tan KB, et al. Application of chitosan and its derivatives as adsorbents for dye removal from water and wastewater: A review. Carbohydrate Polymers. 2014;**113**:115-130

[75] El-Meliegy E, Abu-Elsaad NI, El-Kady AM, Ibrahim MA. Improvement of physico-chemical properties of dextran-chitosan composite scaffolds by addition of nano-hydroxyapatite. Scientific Reports. 2018;**8**(1):12180

[76] Goyal S, Hernández NB, Cochran EW. An update on the future prospects of glycerol polymers. Polymer International. 2021;**70**(7):911-917

[77] Franssen O, Vandervennet L, Roders P, Hennink WE. Degradable dextran hydrogels: Controlled release of a model protein from cylinders and microspheres. Journal of Controlled Release: Official Journal of the Controlled Release Society. 1999;**60**(2-3):211-221

[78] Liebert T, Hornig S, Hesse S, Heinze T. Nanoparticles on the basis of highly functionalized dextrans. Journal of the American Chemical Society. 2005;**127**(30):10484-10485

[79] Long L-x, Yuan X-b, Chang J, Zhang Z-h, Gu M-q, Song T-t, et al. Self-assembly of polylactic acid and cholesterol-modified dextran into hollow nanocapsules. Carbohydrate Polymers. 2012;**87**(4):2630-2637

[80] Sun G, Chu CC. Self-assembly of chemically engineered hydrophilic dextran into microscopic tubules. ACS Nano. 2009;**3**(5):1176-1182

[81] Sun G, Shen YI, Ho CC, Kusuma S, Gerecht S. Functional groups affect physical and biological properties of dextran-based hydrogels. Journal of Biomedical Materials Research Part A. 2010;**93**(3):1080-1090

[82] Sun G, Chen FA, Chu CC. Effects of precursor and cross-linking parameters on the properties of dextran-allyl isocyanate-ethylamine/poly(ethylene glycol diacrylate) biodegradable hydrogels and their release of ovalbumin. Journal of Biomaterials Science Polymer Edition. 2009;**20**(14):2003-2022

[83] Askari M, Fisher C, Weniger FG, Bidic S, Lee WP. Anticoagulation therapy in microsurgery: A review. The Journal of Hand Surgery. 2006;**31**(5):836-846

[84] Booi DI. Perioperative fluid overload increases anastomosis thrombosis in the free TRAM flap used for breast reconstruction. European Journal of Plastic Surgery. 2011;**34**(2):81-86

[85] Gombocz K, Beledi A, Alotti N, Kecskés G, Gábor V, Bogár L, et al. Influence of dextran-70 on systemic inflammatory response and myocardial ischaemia-reperfusion following cardiac operations. Critical Care (London, England). 2007;**11**(4):R87

[86] Steinbauer M, Harris AG, Messmer K. Effects of dextran on microvascular ischemia-reperfusion injury in striated muscle. The American Journal of Physiology. 1997;**272**(4 Pt 2):H1710-H1716

[87] Sakamaki F, Hoffmann H, Müller C, Dienemann H, Messmer K, Schildberg FW. Reduced lipid peroxidation and ischemia-reperfusion injury after lung transplantation using low-potassium dextran solution for lung preservation. American Journal of Respiratory and Critical Care Medicine. 1997;**156**(4 Pt 1):1073-1081

[88] Banz Y, Gajanayake T, Matozan K, Yang Z, Rieben R. Dextran sulfate modulates MAP kinase signaling and reduces endothelial injury in a rat aortic clamping model. Journal of Vascular Surgery. 2009;**50**(1):161-170

[89] Weng L, Romanov A, Rooney J, Chen W. Non-cytotoxic, in situ gelable

hydrogels composed of N-carboxyethyl chitosan and oxidized dextran. Biomaterials. 2008;**29**(29):3905-3913

[90] Draye JP, Delaey B, Van de Voorde A, Van Den Bulcke A, De Reu B, Schacht E. In vitro and in vivo biocompatibility of dextran dialdehyde cross-linked gelatin hydrogel films. Biomaterials. 1998;**19**(18):1677-1687

[91] Fan L, Lin C, Zhao P, Wen X, Li G. An injectable bioorthogonal dextran hydrogel for enhanced chondrogenesis of primary stem cells. Tissue Engineering Part C: Methods. 2018;**24**(9):504-513

[92] Campos FS, Cassimiro DL, Crespi MS, Almeida AE, Gremião MPD. Preparation and characterisation of Dextran-70 hydrogel for controlled release of praziquantel. Brazilian Journal of Pharmaceutical Sciences. 2013;**49**:75-83

[93] Zhang M, Huang Y, Pan W, Tong X, Zeng Q, Su T, et al. Polydopamine-incorporated dextran hydrogel drug carrier with tailorable structure for wound healing. Carbohydrate Polymers. 2021;**253**:117213

[94] Ahmad S, Ahmad M, Manzoor K, Purwar R, Ikram S. A review on latest innovations in natural gums based hydrogels: Preparations & applications. International Journal of Biological Macromolecules. 2019;**136**:870-890

[95] Mahmood H, Khan IU, Asif M, Khan RU, Asghar S, Khalid I, et al. In vitro and in vivo evaluation of gellan gum hydrogel films: Assessing the co impact of therapeutic oils and ofloxacin on wound healing. International Journal of Biological Macromolecules. 2021;**166**:483-495

[96] Palumbo FS, Federico S, Pitarresi G, Fiorica C, Giammona G. Gellan gum-based delivery systems of therapeutic agents and cells. Carbohydrate Polymers. 2020;**229**:115430

[97] Padil VVT, Cheong JY, Kp A, Makvandi P, Zare EN, Torres-Mendieta R, et al. Electrospun fibers based on carbohydrate gum polymers and their multifaceted applications. Carbohydrate Polymers. 2020;**247**:116705

[98] Osmałek TZ, Froelich A, Soból M, Milanowski B, Skotnicki M, Kunstman P, et al. Gellan gum macrobeads loaded with naproxen: The impact of various naturally derived polymers on pH-dependent behavior. Journal of Biomaterials Applications. 2018;**33**(1):140-155

[99] Choi JH, Park A, Lee W, Youn J, Rim MA, Kim W, et al. Preparation and characterization of an injectable dexamethasone-cyclodextrin complexes-loaded gellan gum hydrogel for cartilage tissue engineering. Journal of Controlled Release: Official Journal of the Controlled Release Society. 2020;**327**:747-765

[100] Kim WK, Choi JH, Shin ME, Kim JW, Kim PY, Kim N, et al. Evaluation of cartilage regeneration of chondrocyte encapsulated gellan gum-based hyaluronic acid blended hydrogel. International Journal of Biological Macromolecules. 2019;**141**:51-59

[101] Kim W, Choi JH, Kim P, Youn J, Song JE, Motta A, et al. Preparation and evaluation of gellan gum hydrogel reinforced with silk fibers with enhanced mechanical and biological properties for cartilage tissue engineering. Journal of Tissue Engineering and Regenerative Medicine. 2021;**15**(11):936-947

[102] Vuornos K, Huhtala H, Kääriäinen M, Kuismanen K, Hupa L,

Kellomäki M, et al. Bioactive glass ions for in vitro osteogenesis and microvascularization in gellan gum-collagen hydrogels. Journal of Biomedical Materials Research Part B, Applied Biomaterials. 2020;**108**(4):1332-1342

[103] Lopez-Heredia MA, Łapa A, Reczyńska K, Pietryga K, Balcaen L, Mendes AC, et al. Mineralization of gellan gum hydrogels with calcium and magnesium carbonates by alternate soaking in solutions of calcium/magnesium and carbonate ion solutions. Journal of Tissue Engineering and Regenerative Medicine. 2018;**12**(8):1825-1834

[104] Bastos AR, da Silva LP, Maia FR, Pina S, Rodrigues T, Sousa F, et al. Lactoferrin-hydroxyapatite containing spongy-like hydrogels for bone tissue engineering. Materials (Basel, Switzerland). 2019;**12**(13):2074

[105] Pereira D, Canadas R, Silva-Correia J, da Silva MA, Oliveira M, Dias I, et al. Injectable gellan-gum/hydroxyapatite-based bilayered hydrogel composites for osteochondral tissue regeneration. Applied Materials Today. 2018;**12**:309-321

[106] Gomes ED, Ghosh B, Lima R, Goulão M, Moreira-Gomes T, Martins-Macedo J, et al. Combination of a gellan gum-based hydrogel with cell therapy for the treatment of cervical spinal cord injury. Frontiers in Bioengineering and Biotechnology. 2020;**8**:984

[107] Rezvani Ghomi E, Khalili S, Nouri Khorasani S, Esmaeely Neisiany R, Ramakrishna S. Wound dressings: Current advances and future directions. Journal of Applied Polymer Science. 2019;**136**(27):47738

[108] Shukla R, Kashaw SK, Jain AP, Lodhi S. Fabrication of apigenin loaded gellan gum-chitosan hydrogels (GGCH-HGs) for effective diabetic wound healing. International Journal of Biological Macromolecules. 2016;**91**:1110-1119

[109] Si K, Jeon GY, Kim SE, Choe SH, Kim SJ, Seo JS, et al. Injectable hydrogel based on gellan gum/silk sericin for application as a retinal pigment epithelium cell carrier. ACS Omega. 2022;**7**(45):41331-41340

[110] Bacelar AH, Silva-Correia J, Oliveira JM, Reis RL. Recent progress in gellan gum hydrogels provided by functionalization strategies. Journal of Materials Chemistry B. 2016;**4**(37):6164-6174

[111] Kleinman HK, Martin GR. Matrigel: Basement membrane matrix with biological activity. Seminars in Cancer Biology. 2005;**15**(5):378-386

[112] Corning Incorporated Life Sciences. Corning Matrigel Matrix Frequently Asked Questions. 2019. Available from: https://www.corningcom/catalog/cls/documents/faqs/CLS-DL-CC-026-A4pdf

[113] Hughes CS, Postovit LM, Lajoie GA. Matrigel: A complex protein mixture required for optimal growth of cell culture. Proteomics. 2010;**10**(9):1886-1890

[114] Vukicevic S, Kleinman HK, Luyten FP, Roberts AB, Roche NS, Reddi AH. Identification of multiple active growth factors in basement membrane Matrigel suggests caution in interpretation of cellular activity related to extracellular matrix components. Experimental Cell Research. 1992;**202**(1):1-8

[115] Talbot NC, Caperna TJ. Proteome array identification of bioactive soluble proteins/peptides in Matrigel: Relevance

to stem cell responses. Cytotechnology. 2015;**67**:873-883

[116] Gillette KM, Forbes K, Sehgal I. Detection of matrix metalloproteinases (MMP), tissue inhibitor of metalloproteinase-2, urokinase and plasminogen activator inhibitor-1 within matrigel and growth factor-reduced matrigel basement membrane. Tumori Journal. 2003;**89**(4):421-425

[117] Xu C, Inokuma MS, Denham J, Golds K, Kundu P, Gold JD, et al. Feeder-free growth of undifferentiated human embryonic stem cells. Nature Biotechnology. 2001;**19**(10):971-974

[118] Qian L, Saltzman WM. Improving the expansion and neuronal differentiation of mesenchymal stem cells through culture surface modification. Biomaterials. 2004;**25**(7-8):1331-1337

[119] Lee S-W, Lee HJ, Hwang HS, Ko K, Han DW, Ko K. Optimization of Matrigel-based culture for expansion of neural stem cells. Animal Cells and Systems. 2015;**19**(3):175-180

[120] Laflamme MA, Chen KY, Naumova AV, Muskheli V, Fugate JA, Dupras SK, et al. Cardiomyocytes derived from human embryonic stem cells in pro-survival factors enhance function of infarcted rat hearts. Nature Biotechnology. 2007;**25**(9):1015-1024

[121] Ponce ML. Tube formation: An in vitro matrigel angiogenesis assay. In: Angiogenesis Protocols. United States: Springer; 2009. pp. 183-188

[122] Ponce ML. In vitro matrigel angiogenesis assays. Angiogenesis Protocols. 2001;**46**:205-209

[123] Hocevar SE, Liu L, Duncan RK. Matrigel is required for efficient differentiation of isolated, stem cell-derived otic vesicles into inner ear organoids. Stem Cell Research. 2021;**53**:102295

[124] Lugassy C, Zadran S, Bentolila L, Wadehra M, Prakash R, Carmichael S, et al. Angiotropism, pericytic mimicry and extravascular migratory metastasis in melanoma: An alternative to intravascular cancer dissemination. Cancer Microenviron. 2014;**7**(3):139, 101007/s1230.2014:7-014-152

[125] Arnaoutova I, George J, Kleinman HK, Benton G. Basement membrane matrix (BME) has multiple uses with stem cells. Stem Cell Reviews and Reports. 2012;**8**:163-169

[126] Gehrcke M, Martins CC, de Bastos BT, da Rosa LS, Luchese C, Wilhelm EA, et al. Novel pullulan/gellan gum bilayer film as a vehicle for silibinin-loaded nanocapsules in the topical treatment of atopic dermatitis. Pharmaceutics. 2022;**14**(11):2352

[127] Benton G, Kleinman HK, George J, Arnaoutova I. Multiple uses of basement membrane-like matrix (BME/Matrigel) in vitro and in vivo with cancer cells. International Journal of Cancer. 2011;**128**(8):1751-1757

[128] Scherzer T, Beckert A, Langguth H, Rummel S, Mehnert R. Electron beam curing of methacrylated gelatin. I. Dependence of the degree of crosslinking on the irradiation dose. Journal of Applied Polymer Science. 1997;**63**(10):1303-1312

[129] Shie M-Y, Lee J-J, Ho C-C, Yen S-Y, Ng HY, Chen Y-W. Effects of gelatin methacrylate bio-ink concentration on mechano-physical properties and human dermal fibroblast behavior. Polymers. 2020;**12**(9):1930

[130] Mei Q, Rao J, Bei HP, Liu Y, Zhao X. 3D bioprinting photo-crosslinkable hydrogels for bone and cartilage repair. International Journal of Bioprinting. 2021;**7**(3):367

[131] Choi YH, Kim S-H, Kim I-S, Kim K, Kwon SK, Hwang NS. Gelatin-based micro-hydrogel carrying genetically engineered human endothelial cells for neovascularization. Acta Biomaterialia. 2019;**95**:285-296

[132] Lindberg GC, Lim KS, Soliman BG, Nguyen A, Hooper GJ, Narayan RJ, et al. Biological function following radical photo-polymerization of biomedical polymers and surrounding tissues: Design considerations and cellular risk factors. Applied Physics Reviews. 2021;**8**(1):011301

[133] Liu T, Weng W, Zhang Y, Sun X, Yang H. Applications of gelatin methacryloyl (GelMA) hydrogels in microfluidic technique-assisted tissue engineering. Molecules. 2020;**25**(22):5305

[134] Rahali K, Ben Messaoud G, Kahn CJ, Sanchez-Gonzalez L, Kaci M, Cleymand F, et al. Synthesis and characterization of nanofunctionalized gelatin methacrylate hydrogels. International Journal of Molecular Sciences. 2017;**18**(12):2675

[135] Wu J, Li G, Ye T, Lu G, Li R, Deng L, et al. Stem cell-laden injectable hydrogel microspheres for cancellous bone regeneration. Chemical Engineering Journal. 2020;**393**:124715

[136] Fedorovich NE, Oudshoorn MH, van Geemen D, Hennink WE, Alblas J, Dhert WJ. The effect of photopolymerization on stem cells embedded in hydrogels. Biomaterials. 2009;**30**(3):344-353

[137] Sun M, Sun X, Wang Z, Guo S, Yu G, Yang H. Synthesis and properties of gelatin methacryloyl (GelMA) hydrogels and their recent applications in load-bearing tissue. Polymers. 2018;**10**(11):1290

[138] Wang Y, Zhang R, Qin W, Dai J, Zhang Q, Lee K, et al. Physicochemical properties of gelatin films containing tea polyphenol-loaded chitosan nanoparticles generated by electrospray. Materials & Design. 2020;**185**:108277

[139] Liu J, Li L, Suo H, Yan M, Yin J, Fu J. 3D printing of biomimetic multi-layered GelMA/nHA scaffold for osteochondral defect repair. Materials & Design. 2019;**171**:107708

[140] Alge DL, Anseth KS. Lighting the way. Nature Materials. 2013;**12**(11):950-952

[141] Luo Y, Shoichet MS. A photolabile hydrogel for guided three-dimensional cell growth and migration. Nature Materials. 2004;**3**(4):249-253

[142] West JL. Customized cell microenvironments. Nature Materials. 2011;**10**(10):727-729

[143] Wu T, Gao YY, Su J, Tang XN, Chen Q, Ma LW, et al. Three-dimensional bioprinting of artificial ovaries by an extrusion-based method using gelatin-methacryloyl bioink. Climacteric: The Journal of the International Menopause Society. 2022;**25**(2):170-178

[144] Luo Z, Sun W, Fang J, Lee K, Li S, Gu Z, et al. Biodegradable gelatin methacryloyl microneedles for transdermal drug delivery. Advanced Healthcare Materials. 2019;**8**(3):1801054

[145] Lee K, Xue Y, Lee J, Kim H, Liu Y, Tebon P, et al. A patch of detachable hybrid microneedle depot for localized

delivery of mesenchymal stem cells in regeneration therapy. Advanced Functional Materials. 2020;**30**:2000086

[146] Qu M, Kim H-J, Zhou X, Wang C, Jiang X, Zhu J, et al. Biodegradable microneedle patch for transdermal gene delivery. Nanoscale. 2020;**12**(32):16724-16729

[147] Caudill C, Perry JL, Iliadis K, Tessema AT, Lee BJ, Mecham BS, et al. Transdermal vaccination via 3D-printed microneedles induces potent humoral and cellular immunity. Proceedings of the National Academy of Sciences. 2021;**118**(39):e2102595118

[148] Chi J, Zhang X, Chen C, Shao C, Zhao Y, Wang Y. Antibacterial and angiogenic chitosan microneedle array patch for promoting wound healing. Bioactive Materials. 2020;**5**(2):253-259

[149] Barnum L, Samandari M, Schmidt TA, Tamayol A. Microneedle arrays for the treatment of chronic wounds. Expert Opinion on Drug Delivery. 2020;**17**(12):1767-1780

[150] Zhang X, Hasani-Sadrabadi MM, Zarubova J, Dashtimighadam E, Haghniaz R, Khademhosseini A, et al. Immunomodulatory microneedle patch for periodontal tissue regeneration. Matter. 2022;**5**(2):666-682

[151] Su G, Li G, Wang W, Xu L. Application Prospect and preliminary exploration of GelMA in corneal stroma regeneration. Polymers. 2022;**14**(19):4227

[152] Man K, Barroso IA, Brunet MY, Peacock B, Federici AS, Hoey DA, et al. Controlled release of epigenetically-enhanced extracellular vesicles from a GelMA/Nanoclay composite hydrogel to promote bone repair. International Journal of Molecular Sciences. 2022;**23**(2):832

[153] Hashemi S, Mondal D, Montesano J, Willett TL. Effects of biopolymer functionalization and nanohydroxyapatite heat treatment on the tensile and thermomechanical properties of bone-inspired 3D printable nanocomposite biomaterials. Materials & Design. 2023;**225**:111587

[154] Zhang L, Li T, Yu Y, Shi K, Bei Z, Qian Y, et al. An injectable conductive hydrogel restores electrical transmission at myocardial infarct site to preserve cardiac function and enhance repair. Bioactive Materials. 2023;**20**:339-354

[155] Wu T, Liu W. Functional hydrogels for the treatment of myocardial infarction. NPG Asia Materials. 2022;**14**(1):9

[156] Haghniaz R, Kim H-J, Montazerian H, Baidya A, Tavafoghi M, Chen Y, et al. Tissue adhesive hemostatic microneedle arrays for rapid hemorrhage treatment. Bioactive Materials. 2023;**23**:314-327

[157] Li S, Sun J, Yang J, Yang Y, Ding H, Yu B, et al. Gelatin methacryloyl (GelMA) loaded with concentrated hypoxic pretreated adipose-derived mesenchymal stem cells(ADSCs) conditioned medium promotes wound healing and vascular regeneration in aged skin. Biomaterials Research. 2023;**27**:11

[158] Zhao W, Zhu J, Hang J, Zeng W. Biomaterials to promote vascularization in tissue engineering organs and ischemic fibrotic diseases. MedComm–Biomaterials and Applications. 2022;**1**(1):e16

[159] Rahmani S, Maroufkhani M, Mohammadzadeh-Komuleh S, Khoubi-Arani Z. Chapter 7- Polymer nanocomposites for biomedical applications. In: Barhoum A, Jeevanandam J, Danquah MK, editors.

Fundamentals of Bionanomaterials. Amsterdam, Netherlands: Elsevier; 2022. pp. 175-215

[160] Lee JM, Ma WC, Yeong WY. Chapter 7 - Hydrogels for bioprinting. In: Zhang LG, Fisher JP, Leong KW, editors. 3D Bioprinting and Nanotechnology in Tissue Engineering and Regenerative Medicine. 2nd ed. United States: Academic Press; 2022. pp. 185-211

[161] Hamid ZA, Lim K. Evaluation of UV-crosslinked poly (ethylene glycol) diacrylate/poly (dimethylsiloxane) dimethacrylate hydrogel: Properties for tissue engineering application. Procedia Chemistry. 2016;**19**:410-418

[162] Peppas NA, Keys KB, Torres-Lugo M, Lowman AM. Poly (ethylene glycol)-containing hydrogels in drug delivery. Journal of Controlled Release. 1999;**62**(1-2):81-87

[163] Zhu J. Bioactive modification of poly (ethylene glycol) hydrogels for tissue engineering. Biomaterials. 2010;**31**(17):4639-4656

[164] Skardal A, Zhang J, Prestwich GD. Bioprinting vessel-like constructs using hyaluronan hydrogels crosslinked with tetrahedral polyethylene glycol tetracrylates. Biomaterials. 2010;**31**(24):6173-6181

[165] Kamaraj M, Sreevani G, Prabusankar G, Rath SN. Mechanically tunable photo-cross-linkable bioinks for osteogenic differentiation of MSCs in 3D bioprinted constructs. Materials Science and Engineering: C. 2021;**131**:112478

[166] Cui X, Breitenkamp K, Finn M, Lotz M, D'Lima DD. Direct human cartilage repair using three-dimensional bioprinting technology. Tissue Engineering Part A. 2012;**18**(11-12):1304-1312

[167] Laurienzo P, Malinconico M, Motta A, Vicinanza A. Synthesis and characterization of a novel alginate–poly (ethylene glycol) graft copolymer. Carbohydrate Polymers. 2005;**62**(3):274-282

[168] Jeon O, Samorezov JE, Alsberg E. Single and dual crosslinked oxidized methacrylated alginate/PEG hydrogels for bioadhesive applications. Acta Biomaterialia. 2014;**10**(1):47-55

[169] Olthof MG, Kempen DH, Liu X, Dadsetan M, Tryfonidou MA, Yaszemski MJ, et al. Bone morphogenetic protein-2 release profile modulates bone formation in phosphorylated hydrogel. Journal of Tissue Engineering and Regenerative Medicine. 2018;**12**(6):1339-1351

[170] Liu X, Fundora KA, Zhou Z, Miller AL, Lu L. Composite hydrogel embedded with porous microspheres for long-term pH-sensitive drug delivery. Tissue Engineering Part A. 2019;**25**(3-4):172-182

[171] Chimene D, Kaunas R, Gaharwar AK. Hydrogel bioink reinforcement for additive manufacturing: A focused review of emerging strategies. Advanced Materials. 2020;**32**(1):1902026

[172] Deshmukh K, Basheer Ahamed M, Deshmukh RR, Khadheer Pasha SK, Bhagat PR, Chidambaram K. 3 - Biopolymer composites with high dielectric performance: Interface engineering. In: Sadasivuni KK, Ponnamma D, Kim J, Cabibihan JJ, AlMaadeed MA, editors. Biopolymer Composites in Electronics. Amsterdam, Netherlands: Elsevier; 2017. pp. 27-128

[173] Woodruff MA, Hutmacher DW. The return of a forgotten polymer—Polycaprolactone in the 21st century. Progress in Polymer Science. 2010;**35**(10):1217-1256

[174] Más Estellés J, Vidaurre A, Meseguer Dueñas JM, Castilla CI. Physical characterization of polycaprolactone scaffolds. Journal of Materials Science: Materials in Medicine. 2008;**19**:189-195

[175] Alix S, Mahieu A, Terrie C, Soulestin J, Gerault E, Feuilloley MGJ, et al. Active pseudo-multilayered films from polycaprolactone and starch based matrix for food-packaging applications. European Polymer Journal. 2013;**49**(6):1234-1242

[176] Das GS, Rao GH, Wilson RF, Chandy T. Colchicine encapsulation within poly (ethylene glycol)-coated poly (lactic acid)/poly (?-caprolactone) microspheres-controlled release studies. Drug Delivery. 2000;**7**(3):129-138

[177] Dordunoo SK, Jackson JK, Arsenault LA, Oktaba AMC, Hunter WL, Burt HM. Taxol encapsulation in poly (ε-caprolactone) microspheres. Cancer Chemotherapy and Pharmacology. 1995;**36**:279-282

[178] Chang R-K, Price JC, Whitworth CW. Dissolution characteristics of polycaprolactone-polylactide microspheres of chlorpromazine. Drug Development and Industrial Pharmacy. 1986;**12**(14):2355-2380

[179] Carreras N, Acuña V, Martí M, Lis M. Drug release system of ibuprofen in PCL-microspheres. Colloid and Polymer Science. 2013;**291**(1):157-165

[180] Giunchedi P, Conti B, Maggi L, Conte U. Cellulose acetate butyrate and polycaprolactone for ketoprofen spray-dried microsphere preparation. Journal of Microencapsulation. 1994;**11**(4):381-393

[181] Perez MH, Zinutti C, Lamprecht A, Ubrich N, Astier A, Hoffman M, et al. The preparation and evaluation of poly (ϵ-caprolactone) microparticles containing both a lipophilic and a hydrophilic drug. Journal of Controlled Release. 2000;**65**(3):429-438

[182] Radhakrishnan S, Nagarajan S, Belaid H, Farha C, Iatsunskyi I, Coy E, et al. Fabrication of 3D printed antimicrobial polycaprolactone scaffolds for tissue engineering applications. Materials Science and Engineering: C. 2021;**118**:111525

[183] Mogoşanu GD, Grumezescu AM. Natural and synthetic polymers for wounds and burns dressing. International Journal of Pharmaceutics. 2014;**463**(2):127-136

[184] Zopf DA, Flanagan CL, Wheeler M, Hollister SJ, Green GE. Treatment of severe porcine tracheomalacia with a 3-dimensionally printed, bioresorbable, external airway splint. JAMA Otolaryngology–Head & Neck Surgery. 2014;**140**(1):66-71

[185] Kankala RK, Zhu K, Li J, Wang C-S, Wang S-B, Chen A-Z. Fabrication of arbitrary 3D components in cardiac surgery: From macro-, micro-to nanoscale. Biofabrication. 2017;**9**(3):032002

[186] Hendow EK, Guhmann P, Wright B, Sofokleous P, Parmar N, Day RM. Biomaterials for hollow organ tissue engineering. Fibrogenesis & Tissue Repair. 2016;**9**(1):1-7

[187] Rasperini G, Pilipchuk S, Flanagan C, Park C, Pagni G, Hollister S, et al. 3D-printed bioresorbable scaffold

for periodontal repair. Journal of Dental Research. 2015;**94**(9_suppl):153S-157S

[188] Hiremath P, Nuguru K, Agrahari V. Chapter 8 - Material attributes and their impact on wet granulation process performance. In: Narang AS, Badawy SIF, editors. Handbook of Pharmaceutical Wet Granulation. United States: Academic Press; 2019. pp. 263-315

[189] Folttmann H, Quadir A. Polyvinylpyrrolidone (PVP)—One of the most widely used excipients in pharmaceuticals: An overview. Drug Delivery Technology. 2008;**8**(6):22-27

[190] Bothiraja C, Shinde MB, Rajalakshmi S, Pawar AP. Evaluation of molecular pharmaceutical and in-vivo properties of spray-dried isolated andrographolide—PVP. Journal of Pharmacy and Pharmacology. 2009;**61**(11):1465-1472

[191] Rasekh M, Karavasili C, Soong YL, Bouropoulos N, Morris M, Armitage D, et al. Electrospun PVP–indomethacin constituents for transdermal dressings and drug delivery devices. International Journal of Pharmaceutics. 2014;**473**(1-2):95-104

[192] del Consuelo ID, Falson F, Guy RH, Jacques Y. Ex vivo evaluation of bioadhesive films for buccal delivery of fentanyl. Journal of Controlled Release. 2007;**122**(2):135-140

[193] Zhang L, Liang Y, Meng L, Wang C. Characterization of complexation of PVP copolymer with DNA. Polymers for Advanced Technologies. 2009;**20**(4):410-415

[194] Song Y, Zhang T, Song X, Zhang L, Zhang C, Xing J, et al. Polycations with excellent gene transfection ability based on PVP-g-PDMAEMA with random coil and micelle structures as non-viral gene vectors. Journal of Materials Chemistry B. 2015;**3**(5):911-918

[195] Saxena A, Mozumdar S, Johri AK. Ultra-low sized cross-linked polyvinylpyrrolidone nanoparticles as non-viral vectors for in vivo gene delivery. Biomaterials. 2006;**27**(32):5596-5602

[196] Benamer S, Mahlous M, Boukrif A, Mansouri B, Youcef SL. Synthesis and characterisation of hydrogels based on poly(vinyl pyrrolidone). Nuclear Instruments and Methods in Physics Research Section B: Beam Interactions with Materials and Atoms. 2006;**248**(2):284-290

[197] Daniele MA, Adams AA, Naciri J, North SH, Ligler FS. Interpenetrating networks based on gelatin methacrylamide and PEG formed using concurrent thiol click chemistries for hydrogel tissue engineering scaffolds. Biomaterials. 2014;**35**(6):1845-1856

[198] An J, Teoh JEM, Suntornnond R, Chua CK. Design and 3D printing of scaffolds and tissues. Engineering. 2015;**1**(2):261-268

[199] Yoshida K, Sakurai Y, Kawahara S, Takeda T, Ishikawa T, Murakami T, et al. Anaphylaxis to polyvinylpyrrolidone in povidone-iodine for impetigo contagiosum in a boy with atopic dermatitis. International Archives of Allergy and Immunology. 2008;**146**(2):169-173

[200] Adachi A, Fukunaga A, Hayashi K, Kunisada M, Horikawa T. Anaphylaxis to polyvinylpyrrolidone after vaginal application of povidone-iodine. Contact Dermatitis. 2003;**48**(3):133-136

[201] Rönnau AC, Wulferink M, Gleichmann E, Unver E, Ruzicka T, Krutmann J, et al. Anaphylaxis to

polyvinylpyrrolidone in an analgesic preparation. British Journal of Dermatology. 2000;**143**(5):1055-1058

[202] York KK, Miller S, Gaster RN, Burstein NL. Polyvinylpyrrolidone iodine: Corneal toxicology and epithelial healing in a rabbit model. Journal of Ocular Pharmacology and Therapeutics. 1988;**4**(4):351-358

[203] Adsul MG, Varma AJ, Gokhale DV. Lactic acid production from waste sugarcane bagasse derived cellulose. Green Chemistry. 2007;**9**(1):58-62

[204] Singhvi M, Joshi D, Adsul M, Varma A, Gokhale D. d-(−)-Lactic acid production from cellobiose and cellulose by lactobacillus lactis mutant RM2-2 4. Green Chemistry. 2010;**12**(6):1106-1109

[205] Datta R, Tsai S-P, Bonsignore P, Moon S-H, Frank JR. Technological and economic potential of poly (lactic acid) and lactic acid derivatives. FEMS Microbiology Reviews. 1995;**16**(2-3):221-231

[206] Gupta B, Revagade N, Hilborn J. Poly (lactic acid) fiber: An overview. Progress in Polymer Science. 2007;**32**(4):455-482

[207] Graupner N, Herrmann AS, Müssig J. Natural and man-made cellulose fibre-reinforced poly (lactic acid)(PLA) composites: An overview about mechanical characteristics and application areas. Composites Part A: Applied Science and Manufacturing. 2009;**40**(6-7):810-821

[208] Dutkiewicz S, Grochowska-Łapienis D, Tomaszewski W. Synthesis of poly (L (+) lactic acid) by polycondensation method in solution. Fibres & Textiles in Eastern Europe. 2003;**4**(43):66-70

[209] Terzopoulou Z, Zamboulis A, Bikiaris DN, Valera MA, Mangas A. Synthesis, properties, and enzymatic hydrolysis of poly(lactic acid)-co-poly(propylene adipate) block copolymers prepared by reactive extrusion. Polymers. 2021;**13**(23):4121

[210] Chen G, Ushida T, Tateishi T. Scaffold design for tissue engineering. Macromolecular Bioscience. 2002;**2**(2):67-77

[211] Lasprilla AJR, Martinez GAR, Lunelli BH, Jardini AL, Filho RM. Poly-lactic acid synthesis for application in biomedical devices — A review. Biotechnology Advances. 2012;**30**(1):321-328

[212] Casalini T, Rossi F, Castrovinci A, Perale G. A perspective on polylactic acid-based polymers use for nanoparticles synthesis and applications. Frontiers in Bioengineering and Biotechnology. 2019;**7**:259

[213] Callahan LAS, Xie S, Barker IA, Zheng J, Reneker DH, Dove AP, et al. Directed differentiation and neurite extension of mouse embryonic stem cell on aligned poly (lactide) nanofibers functionalized with YIGSR peptide. Biomaterials. 2013;**34**(36):9089-9095

[214] Kurobe H, Maxfield MW, Tara S, Rocco KA, Bagi PS, Yi T, et al. Development of small diameter nanofiber tissue engineered arterial grafts. PLoS One. 2015;**10**(4):e0120328

[215] Mohiti-Asli M, Pourdeyhimi B, Loboa EG. Skin tissue engineering for the infected wound site: Biodegradable PLA nanofibers and a novel approach for silver ion release evaluated in a 3D coculture system of keratinocytes and Staphylococcus aureus. Tissue Engineering Part C: Methods. 2014;**20**(10):790-797

[216] Raquez J-M, Habibi Y, Murariu M, Dubois P. Polylactide (PLA)-based nanocomposites. Progress in Polymer Science. 2013;**38**(10-11):1504-1542

[217] Gentile P, Chiono V, Carmagnola I, Hatton PV. An overview of poly(lactic-co-glycolic) acid (PLGA)-based biomaterials for bone tissue engineering. International Journal of Molecular Sciences. 2014;**15**(3):3640-3659

[218] Gupta A, Kumar V. New emerging trends in synthetic biodegradable polymers–polylactide: A critique. European Polymer Journal. 2007;**43**(10):4053-4074

[219] Mir M, Ahmed N, ur Rehman A. Recent applications of PLGA based nanostructures in drug delivery. Colloids and Surfaces B: Biointerfaces. 2017;**159**:217-231

[220] Passerini N, Craig D. An investigation into the effects of residual water on the glass transition temperature of polylactide microspheres using modulated temperature DSC. Journal of Controlled Release. 2001;**73**(1):111-115

[221] Engineer C, Parikh J, Raval A. Review on hydrolytic degradation behavior of biodegradable polymers from controlled drug delivery system. Trends in Biomaterials & Artificial Organs. 2011;**25**(2):79-85

[222] Makadia HK, Siegel SJ. Poly lactic-co-glycolic acid (PLGA) as biodegradable controlled drug delivery carrier. Polymers. 2011;**3**(3):1377-1397

[223] Sharma S, Parmar A, Kori S, Sandhir R. PLGA-based nanoparticles: A new paradigm in biomedical applications. TrAC Trends in Analytical Chemistry. 2016;**80**:30-40

[224] Alkahtani S, Alarifi S, Albasher G, Al-Zharani M, Aljarba NH, Almarzoug MH, et al. Poly lactic-co-glycolic acid- (PLGA-) loaded nanoformulation of cisplatin as a therapeutic approach for breast cancers. Oxidative Medicine and Cellular Longevity. 2021;**2021**:5834418

[225] Lee PW, Pokorski JK. Poly(lactic-co-glycolic acid) devices: Production and applications for sustained protein delivery. Wiley Interdisciplinary Reviews Nanomedicine and Nanobiotechnology. 2018;**10**(5):e1516

[226] Yu Y, Sun B, Yi C, Mo X. Stem cell homing-based tissue engineering using bioactive materials. Frontiers of Materials Science. 2017;**11**:93-105

[227] Han Y, Li X, Zhang Y, Han Y, Chang F, Ding J. Mesenchymal stem cells for regenerative medicine. Cell. 2019;**8**(8):886

[228] Pourentezari M, Dortaj H, Hashemibeni B, Yadegari M, Shahedi A. An overview of the application of poly(lactic-co-glycolic) acid (PLGA)-based scaffold for drug delivery in cartilage tissue engineering. International Journal of Medical Laboratory. 2021;**8**(2):78-95

Chapter 2

Drug-Delivery Based on Encapsulation for Photodynamic Therapy and Photothermal Therapy

María Paulina Romero and Myriam Alexandra González

Abstract

Photodynamic therapy (PDT) and photothermal therapy (PTT) have shown significant promise in treating cancer and other illnesses. Photosensitizers (PSs) and photothermal agents (PTAs) play crucial roles in PDT and PTT to enhance treatment efficiently. The stability, solubility, and toxicity of PSs and PTAs are the main challenges in improving the effectiveness of these agents. This chapter examines several encapsulation techniques of PS and PTAs agents to reduce these limitations. Liposomes, hydrogels, inorganic particles, metal-organic frameworks, and covalent organic frameworks offer diverse and important strategies for PS, PTAs agents' delivery. These nanocarrier systems offer unique advantages such as high encapsulation efficiency, sustained drug release, enhanced cytotoxicity against cancer cells, and biocompatibility. Moreover, the nanocapsules create a localized drug depot at the target site and present specific benefits tailored to the requirements of different therapeutic applications, making them valuable tools in developing advanced and targeted drug delivery systems for cancer and other diseases. Overall, the diverse strategies for encapsulating photosensitizers offer unique advantages for enhancing the efficacy and safety of PDT and PTT, making them promising candidates for PDT and drug delivery applications.

Keywords: encapsulation, photosensitizers, PDT, PTT, polymers, hydrogels, liposomes, COF, MOF, inorganic nanoparticles, micelles

1. Introduction

Photodynamic therapy (PDT) and photothermal therapy (PTT) are highly effective treatments for cancer, as they can suppress tumors while minimizing damage to surrounding tissue [1] (**Figure 1**). PDT is a precise and focused approach that can be utilized alone or with other conventional therapies, such as chemotherapy, sonotherapy, and photothermal therapy. Unlike conventional therapies, PDT uses light as a targeting mechanism, making it simpler and more direct. The photosensitizer (PS), which converts light energy into therapeutic elements, is the primary component of a

IntechOpen

PDT system [1–3]. Upon exposure to light, these PSs can react with molecular oxygen ($^3O_2$) to produce cytotoxicity (ROS). This process can occur through the production of singlet oxygen ($^1O_2$) via a "type II" pathway or the production of superoxide anion radical ($^1O_2^-$) through a "type I" mechanism involving an electron transfer [4]. In PDT, the efficacy of the treatments relies on PSs to generate ROS (**Figure 2**) [5, 6].

PTT involves administering photothermal chemicals into the body, which accumulate specifically at the tumor site [7, 8]. When exposed to radiation from external light sources, such as near-infrared light, the absorbed light energy is efficiently converted into heat energy, resulting in high local internal temperatures within the tumor cells

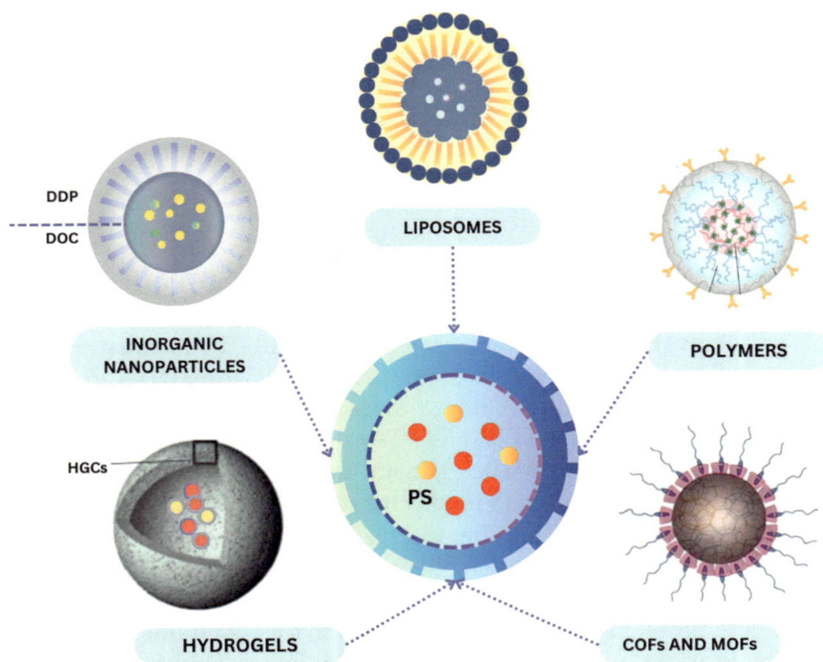

Figure 1.
Principal strategies to encapsulate photosensitizers (PSs) and photothermal agents (PTAs) to PDT and PTT.

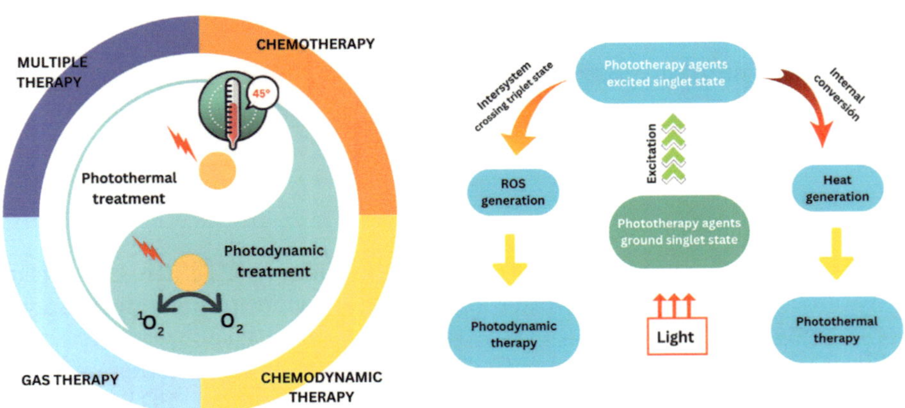

Figure 2.
Mechanism of photodynamic therapy or photothermal therapy activity.

or tissues to induce tumor cell death [8]. PTAs interact with tissue upon exposure to near-infrared (NIR-IR, 650–950 nm) light, leading to induced cell death through an intrinsic apoptotic pathway. PTT has been shown to enhance the expression of pro-apoptotic proteins, increase the activity of caspases (specific cysteine protease), induce mitochondrial apoptosis, promote the entry of more cells into late apoptotic phases, and eliminate superfluous or undesired cells [8–10].

PSs and PTAs have been widely used in cancer therapy. Common PSs such as porphyrins, chlorins, and phthalocyanines have been used in PDT to generate reactive oxygen species (ROS) that can induce cell death. In the same way, PTAs, including a variety of metallic nanostructures (Au-, Ag- and Cu- based nanoparticles), and carbon nanomaterials (graphene and carbon nanotubes) have been used in PTT to generate heat and cause thermal damage to targeted cells [8]. Recent advancements in nanotechnology have led to the development of PSs and PTAs such as porphyrins, phthalocyanines, coumarins, and others. These compounds have been enhanced through encapsulation in polymers, lipids, inorganic nanoparticles, metal-organic frameworks (MOFs), covalent organic frameworks (COFs), and hydrogels [11, 12].

Despite the diverse benefits of PDT and PTT agents, their therapeutic applications are limited by their toxicity and limited solubility in aqueous solutions. Biotechnology and nanotechnology have become crucial instruments in overcoming these challenges by creating novel, more effective methods for drug delivery of photoactive chemicals. Nanoparticles, such as liposomes [13], polymeric nanoparticles [14], and hydrogels [15], have been employed for delivering PSs or PTAs to the target site, allowing for enhanced permeability and retention (EPR) effect and facilitating passive targeting of tumors [8]. Besides, recent developments in nanotechnology have led to the development of new-generation nanoparticles with improved functionalities, such as metal-organic frameworks (MOFs) [16], covalent organic frameworks (COFs) [17], and inorganic nanoparticles [18] that combine the advantages of different materials and nanomaterials with stimuli-responsive features that enable controlled drug release at the target site.

Moreover, encapsulating PSs and PTAs into various carriers such as polymers, liposomes, hydrogels, inorganic particles, MOF, and COF offers diverse and important strategies for PDT and PTT. These strategies can improve the solubility and stability of PSs, reduce their toxicity, increase the agent's accumulation in the tumor site, prolong their retention time, and enhance their specificity and selectivity toward cancer cells [16, 17].

Polymers, such as poly(methacrylic acid) (PMA), poly(lactide-co-glycolide) (PLGA), and poly(vinylcaprolactam) (PVCL), have been utilized to form nanocapsules for the encapsulation of PSs and PTAs [14]. These polymer-based nanocapsules offer advantages such as high encapsulation efficiency, sustained drug release, and enhanced cytotoxicity against cancer cells, making them promising candidates for targeted drug delivery in PDT and PTT [14, 19, 20]. Liposomes are also formulations used for the encapsulation of PDT and PTT agents, offering advantages such as biocompatibility, high drug-loading capacity, and the ability to target specific tissues [13]. Liposomal formulations have shown potential for improving the solubility and stability of photosensitizers and enhancing their accumulation at the target site, thereby improving the efficacy of PDT [13, 21].

In addition to polymers and liposomes, hydrogels have been used to encapsulate photosensitizers, offering advantages such as stimuli-responsive (pH, redox potential, or enzymatic activity) drug release, prolonged circulation time, and a localized drug depot at the target site [13, 20, 22, 23]. Hydrogel-based nanocapsules have demonstrated enhanced biocompatibility and sustained release of PSs and PTAs, making them suitable for PDT and PTT applications. Furthermore, inorganic particles, MOFs, and COFs have also been explored to encapsulate PSs and PTAs, offering unique advantages such as tunable porosity, high surface area, and the ability to encapsulate hydrophobic and hydrophilic drugs [16, 17, 24, 25]. These carriers have shown potential for improving the stability and bioavailability of photosensitizers, as well as enabling controlled drug release, making them promising candidates for PDT and PTT applications. This chapter will explore the use of different carriers to encapsulate PDT and PTT agents.

This chapter aims to discuss the recent advances in drug delivery to PDT and PTT, focusing on encapsulation strategies. The chapter will provide a comprehensive overview of several types of encapsulations, including polymeric core-shell encapsulation, liposomes, MOFs, COFs, inorganic nanoparticles, and hydrogels. These encapsulation methods have shown promise in improving the stability, solubility, and toxicity of PSs and PTAs, thereby enhancing the efficiency of PDT and PTT.

2. Strategies to encapsulate photosensitizers to PDT

2.1 Polymeric encapsulation

Polymeric encapsulation is a drug delivery technique that involves enclosing active substances (PSs and PTAs) within a polymeric matrix or shell [13]. This process creates a protective barrier around the core material, which prevents degradation, improves stability, and controls the release rate of the encapsulated components [26, 27]. Polymeric encapsulation offers several advantages, including targeted drug delivery, enhanced bioavailability, and reduced side effects [28]. There are several methods for encapsulation, including emulsion-solvent evaporation/extraction, coacervation-phase separation, spray drying, interfacial and in situ polymerization, and nanoprecipitation [20]. Each method has its advantages and disadvantages, and the choice of method depends on the specific application and the properties of the active substances.

In addition to the encapsulation methods, the choice of polymer is also essential. Both synthetic and natural polymers are employed in encapsulation. Synthetic polymers, such as polylactic acid (PLA) [29], polyglycolic acid (PGA) [30], and poly(lactic-co-glycolic acid) (PLGA) [31, 32], poly(D,L-lactide)(PDLLA), poly(hydroxybutyrate-co-hydroxyvalerate) (PHBV) [29], and polyethylene glycol-block-poly(D,L-lactide) (PEG-b-PDLA) [32] are widely used due to their biocompatibility, biodegradability, and tunable properties. On the other hand, natural polymers, such as chitosan [33], alginate, and cellulose derivatives, are also used due to their biocompatibility, low toxicity, and availability [34].

Encapsulation of agents for PDT and PTT using various polymers has been studied, demonstrating the potential of this technique in improving the efficacy and safety of these treatments [11, 28]. PLGA and PHBV were combined with lipid-PEG to create polymer-lipid-PEG hybrid nanoparticles to encapsulate the PS 5,10,15,20-tetrakis (4-hydroxy-phenyl)-21H,23H-porphine (pTHPP). The resulting nanocapsules displayed an average size of approximately 88.5 nm for PLGA and 215.0 nm for PHBV. Both systems exhibited high encapsulation efficiency (EE) and excellent loading capacity, with pTHPP molecules showing improved photophysical properties, such as increased singlet oxygen generation and cellular uptake rates in FTC-133 human thyroid carcinoma cell lines. Furthermore, pTHPP-loaded nanocapsules exhibited higher in vitro phototoxicity upon 652 nm laser irradiation compared to free pTHPP, highlighting the effectiveness of these systems for PDT cancer treatment [29].

A similar study used PLGA, PDLLA, and PEG-b-PDLA nanoparticles to encapsulate chloroaluminum phthalocyanine (ClAlPc). These nanoparticles exhibit variable sizes (115–274 nm) and EE between 57 and 96%. PLGA polymers obtained the best EE and size distribution (95% and 178 nm). The negative surface charge (−37 to −59 mV) confers colloidal stability. *In vitro*, experiments demonstrated a phototoxic effect more significant than 80% in human fibroblast cells exposed to low laser light doses (3 J/cm^2 upon 650 nm irradiation laser) and 10 μM of AlClPc [32]. Likewise, Bazylinska et al. proved the properties of PLGA-based nanocarriers mixed with PEG and FA (folic acid) to encapsulate hydrophobic photosensitizer verteporfin (VP) in combination with cisplatin (CisPt) a hydrophilic cancer drug. The diameter of nanocapsules containing VP and CisPt had an average size between 187 and 200 nm, zeta potential from −4 to 0.17 mV, with an efficient assembly of nanocapsules. The nanocapsules showed high stability and release in tumor environment, with excellent photocytotoxicity effect in SKOV-3 cells and enhancing PDT applications in ovarian cancer cells (**Figure 3**) [1].

Additionally, biocompatibility and stability were improved of the andrographolide (ADG) anti-cancer agent through the encapsulation in PLGA-PEG-PLGA amphiphilic triblock copolymer micelles. ADG-loaded PLGA-PEG-PLGA micelles presented a high EE of about 92% and a stable particle size of 124.3 nm. *In vitro*, cytotoxicity tests revealed that ADG-loaded PLGA-PEG-PLGA micelles exerted more potent proliferation inhibition, cell cycle arrest at the G2/M phase, and pro-apoptotic effects in MDA-MB-231 human breast cancer cells than unencapsulated ADG [31, 36]. These studies demonstrate higher encapsulation efficiency with PLGA and PHBV

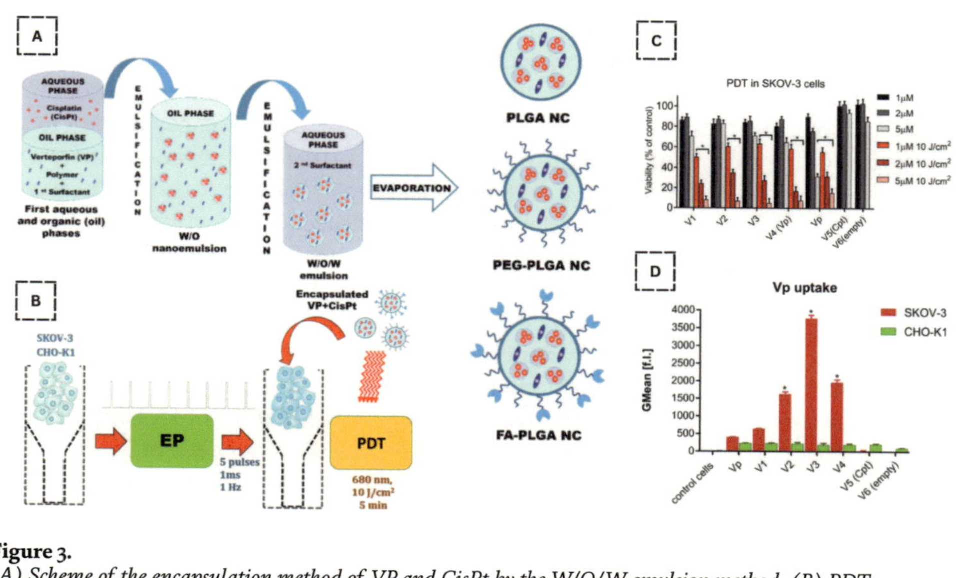

Figure 3.
(A) Scheme of the encapsulation method of VP and CisPt by the W/O/W emulsion method. (B) PDT applications with NCs against human ovarian cancer (SKOV-3) and hamster ovarian control (CHO-K1) cells. (C) Cells viability in SKOV-3 after 24 h of incubation with NCs, after 10 min exposition of 1 μM NCs. (D) Influence of the nanocarrier surface on the encapsulated cargo uptake by SKOV-3 and CHOK1 cells evaluated by flow cytometry analysis. From Bazylińska et al. [35], used under CC-BY 4.0.

encapsulating PSs and anti-cancer agents, with higher phototoxic effects when applied in PDT and PTT treatments [29, 32]. The PLGA-PEG-PLGA micelles and PCL display prolonged photo-induced cytotoxic effects, suggesting their suitability for treating cancer solid tumors and PTT advantages.

Likewise, poly-E-caprolactone (PCL) was used to encapsulate methylene blue (MB) as a PS with an average size of 156 nm and EE of 94.22%. These nanocapsules showed satisfactory ROS generations and higher cytotoxicity under light irradiation (630 nm) [37]. In another study developed by Daskin et al., a docetaxel (DCX) anti-cancer agent was encapsulated in PCL coated with PEG and CS. The nanocapsules, averaging 180–210 nm and 65% of EE, showed significantly higher cytotoxicity at MCF-7 gastrointestinal cancer cells underneath (570 nm green irradiation wave) [38]. These studies demonstrate the advantageous role of encapsulation using a PCL core-shell in enhancing the efficacy and bioavailability of active substances with high stability, biocompatibility, and encapsulation efficiency to application in PDT treatments.

New generations of polymers were also developed. Doshi M et al. explored the use of conducting polymer nanoparticles (CPNPs) composed of poly[2-methoxy-5-(2-ethylhexyl-oxy)-p-phenylenevinylene] (MEH-PPV] for applications in PTT. The study analyzed the level of cellular uptake, cytotoxicity, and the effectiveness of CPNPs in PDT across tumor cells under a 580 nm light dose of 180 J/cm^2 [39]. Similarly, Zhang et al. introduced a novel biodegradable polymer, PPDT, originating from a dihydroxy-containing BODIPY monomer to encapsulate celecoxib (CXB) within PPDT nanoparticles (NPPDT@CXB). Upon an 808 nm laser radiation, NPPDT@CXB instigated an immunogenic cell death (ICD) cascade, initiated anti-tumor immune reactions, and efficiently suppressed primary and metastatic tumors [40]. These studies highlight the potential of these new generations of shells for encapsulating agents for PDT and PTT applications and demonstrate the effectiveness of these systems in improving the therapeutic outcomes of these treatments [39, 40].

The potential of utilizing natural polymers, specifically chitosan, instead of synthetic counterparts for developing nanoparticles in PDT is worth mentioning. Natural polymers offer several advantages, including reduced toxicity and improved biocompatibility relative to synthetic alternatives. In a study focusing on chitosan, functionalized nanostructured lipid carriers (NLCs) were used to encapsulate chloroaluminum phthalocyanine (ClAlPc) to obtain ClAlPc/NLCs nanocapsules, with a mean diameter of 220 nm and zeta potential of +19 mV. The nanocapsules have an EE of 96% of ClAlPc, resulting in enhanced *in vitro* release and penetration of ClAlPc in skin models. Furthermore, ClAlPc/NLCs demonstrated enhanced PDT effectiveness *in vitro*, significantly reducing cell viability upon 660 nm laser irradiation [33, 41]. Similarly, Potara et al. encapsulated PS IR780, an important theranostic agent with lipophilicity limitation, into polymeric Pluronic-F127-chitosan nanocapsules. The targeted nanocapsules exhibited enhanced NIR-laser-induced phototherapeutic performance against ovarian cancer cells compared to free IR780, showcasing the potential of the carrier to deliver lipophilic substances into tumor cells [42].

Additionally, chitosan-based nanoparticles (CNPs) were synthesized as effective drug carriers for PDT. The CNPs exhibited high drug-loading efficiency (>90%) and sustained release profiles *in vitro*. Lee et al. prepared photosensitizer-encapsulated CNPs (PpIX-CNPs) that demonstrated enhanced tumor specificity and increased therapeutic efficacy compared to free PS protoporphyrin IX (PpIX) in tumor-bearing mice [43]. In the same line, Jin et al. encapsulated curcumin (Cur) into CNP core-shell nanoparticles with high encapsulation (85.70%), thermostability, and photostability of PS. These studies demonstrated the potential of chitosan-based nanoparticles as an effective drug delivery system to encapsulate PSs for PDT and PTT applications [44].

Studies have demonstrated the potential of polymeric nanocapsules in improving the efficacy and safety of treatments like PDT and PTT. Encapsulation of PSs and anti-cancer agents within polymers such as PLGA, PHBV, and PCL has shown enhanced encapsulation efficiency, stability, and bioavailability, leading to higher phototoxic effects when applied in cancer treatments. Novel polymer formulations, such as conducting polymer nanoparticles and biodegradable polymers, have also shown promise in enhancing therapeutic outcomes in PDT and PTT applications. In the same line, natural polymers like chitosan have emerged as attractive alternatives to synthetic polymers for developing nanoparticles in PDT due to their reduced toxicity and improved biocompatibility. Chitosan-based nanocapsules have demonstrated high drug-loading efficiency, sustained release profiles, enhanced tumor specificity, and increased therapeutic efficacy in delivering PSs for PDT and PTT applications. These studies highlight the versatility and effectiveness of polymeric nanocapsules in drug delivery for cancer therapy.

2.2 Covalent and metal-organic framework (COFs and MOFs)

Metal-organic frameworks (MOFs) and covalent organic frameworks (COFs) are two porous materials with distinctive features and applications. MOFs are hybrid structures that combine organic ligands and metal ions or clusters to form porous architectures [45]. COFs are crystalline porous materials made of organic building blocks joined by strong covalent bonds to form a stable and well-organized structure [17]. COFs have been used in PDT applications as drug delivery devices because of their large surface area, which makes drug loading and controlled release possible. Additionally, by including photothermal agents within the framework, their adjustable features make them perfect candidates for PTT [27]. Similarly, when MOFs are in the nanoscale (nMOFs), they have high drug-loading ability, biodegradability, enhanced permeability, and retention (EPR). MOFs can be encapsulated as PTA agents in PTT to treat cancer precisely by converting light into heat [17, 45, 46].

Zeolitic imidazolate framework-8 (ZIF-8) is an MOF material that was employed to encapsulate Cu2-XSe, an effective photothermal transduction agent for PTT and PS Indocyanine green (ICG) (ICG@Cu2-XSe-ZIF-8), to treat malignant breast cancer bone tumor. Typical cell damage was avoided because of the ZIF-8 encapsulation, which decreased the accumulation of Cu2-XSe and ICG. ZIF-8 was cleaved to liberate Cu2-XSe in acidic tumor microenvironments. This research showed that the PTT-induced hyperthermia improved the tumor-associated PDT effect, resulting in a synergistic PTT/PDT effect [47]. Similarly, Shao et al. used zeolitic imidazolate frameworks (ZIF-8) and Zr (IV)-porphyritic MOFs to encapsulate doxorubicin (DOX) with a mean diameter of 142 nm. PCN@D/ZIF response in acidic tumor microenvironment enhanced the release of DOX (> 80%). PCN@D/ZIF generated a large amount of singlet oxygen upon light irradiation (640 nm), which increased the therapeutic efficacy of chemo/PDT [24]. In the same line, nMOFs were also used to encapsulate other PSs such as Ce6, CQ, and tirapazamine (TPZ). The use of core-sell nanoparticle@porphyrinic MOFs (UCSs) to encapsulate TPZ (TPZ/UCSs) showed that TPZ/UCSs offered an efficient method to combine NIR light-induced photodynamic therapy and hypoxia-activated chemotherapy with improving cancer treatment both *in vivo* and *in vitro* [48].

Conversely, COFs have become a viable option for encapsulating essential drugs for PDT and PTT. Numerous research has demonstrated how COFs, using creative encapsulation techniques, can augment the effectiveness of these treatment modalities [49]. Li et al. reported the first ROS-responsive dithioketal-linked COF for synergetic chemotherapy and PDT for cancer treatments. This COF demonstrated effective singlet oxygen ($^1O_2$)-responsive dissociation, allowing a more effective combination of chemotherapy and PDT. This work emphasized how drug release mechanisms and COF dissociation are complimentary, offering a viable strategy for dual therapeutic approaches [49].

ICG, a near-infrared photosensitizer, was combined with COF 1,3,5-triformylbenzene and 2,5-bis(methylsulfonyl)-1,4-diaminobenzene) to produce PTT and PDT effects. This COF system demonstrated improved drug loading capacity (53.59%) and encapsulation efficiency (57.75%), along with controlled drug release (80%) when exposed to NIR 808-laser irradiation. The research highlights the potential of COFs for combined PDT/PTT applications by demonstrating significant cytotoxic effects against cancer cells under light irradiation, together with exceptional biocompatibility [50]. Similarly, Ge et al. encapsulated the PS Ce6 and the hypoxia-activated medication TPZ into azo-containing responsive (TA-COFs) to form TA-COF-P@CT with a mean size of 90 nm. These nanocapsules demonstrated a two-step hypoxia-responsive drug release mechanism that inhibited tumor development and eliminated cancer cells in both *in vitro* and *in vivo* upon NIR-650 nm laser irradiation [51]. These studies demonstrated how COFs can respond to drug delivery systems to improve therapeutic outcomes when used with PTT and PDT.

There are several advantages to encapsulating photosensitizers and other agents in COFs and nMOFs. Some of them are enhanced singlet oxygen generation, reduced tumor hypoxia, enhanced biocompatibility, and tailored active targeting capabilities [24, 27, 45, 52]. Because of their huge surface areas, high porosities, and adjustable compositions, the distinct structural advantages of COFs and nMOFs make them ideal for encapsulating and delivering therapeutic compounds for cancer treatment. In addition, the restricted penetration depth and irradiation area of light, which make PDT inappropriate for deep-seated malignancies and metastatic tumors, can be addressed using nMOFs as nanocapsules for PDT and PTT [17, 50–52]. It has been demonstrated that modifying groups to nMOFs enhances PDT's efficacy in laboratory and clinical settings. These alterations have shown remarkable efficacy both *in vivo* and *in vitro*, indicating the possibility of encasing photosensitizers in nMOFs to increase PDT's therapeutic effectiveness.

2.3 Liposomes encapsulation

PDT and PTT can be more successful when photosensitizing drugs are more soluble and bioavailable, which can be achieved through nanoencapsulation. Liposomes (LPs) are produced spontaneously by phospholipids, forming a bilayer that naturally tends to fold in on itself [13, 53]. LPs can be employed as drug carriers because of their lipid composition, which also offers benefits, including reduced toxicity, biodegradability, and biocompatibility [54]. They enable the prolonged release of hydrophilic and hydrophobic compounds and their encapsulation as adaptable drug carriers. Various lipids can make up LPs, such as naturally occurring and manufactured lipids with varying saturation levels, positively charged, negatively charged, or neutral lipids, among many other characteristics that need to be considered during manufacturing [55]. New liposome derivatives, such as ethosomes and transfersomes, provide the vesicles more flexibility and improve permeability [13, 56]. However, ethosomes have a high ethanol content, transfersomes are composed primarily of phospholipids with the addition of an edge activator, such as a surfactant. Comparing these methods to traditional liposomes reveals that they are more flexible, and the penetration enhancer agents claim that the vesicles can reach deeper skin layers [13, 53–55].

Single liposomes and co-encapsulation are the two primary encapsulation techniques investigated to enclose PSs and PTAs. MB and AO have been moderately encapsulated by single liposomes, which contain various lipids and may or may not contain surfactants [57]. According to Pivetta et al. AO and MB were encapsulated in LPs vesicles by the thin-film hydration method to form AO/LPs-NCs and MB/LPs-NCs. The AO had a higher affinity for vesicles with EE, more than 98%, whereas MB exhibited moderate encapsulation (63–86%). MET1 squamous cell carcinoma (SCC) cells treated with AO liposomes showed increased phototoxicity. The nanocapsules demonstrated photosensitivity under blue light (430–490 nm), light frequently employed in PDT (**Figure 4A**) [57, 58]. Similarly, Prathyusha et al. demonstrated ROS generating capacity of Cur natural PS encapsulated into LPs with the same method, obtaining nanocapsules with a mean size of 200 nm and EE of 66.8%. *In vitro* studies in breast cancer cells (MCF-7) demonstrated a sustained release of the Cur, ROS regenerating capabilities, and cytotoxicity under blue light-460 nm irradiation [60].

Conversely, co-encapsulation of two or more photosensitizers in LPs has been studied for PDT and PTT. LPs were used to encapsulate ICG and Ce6 PSs to create a thermosensitive drug delivery system, ICG/LPs-NCs and Lip(Ce6); under NIR irradiation (808 nm), the resultant liposome capsules exhibited good stability and effective inhibition of breast cancer cell growth, with an average diameter ranging from 71 to 127 nm [61–63]. **Figure 4B-E**, shows the principal results from the study. Based on these results, it can be demonstrated that these nanocapsules, with a size

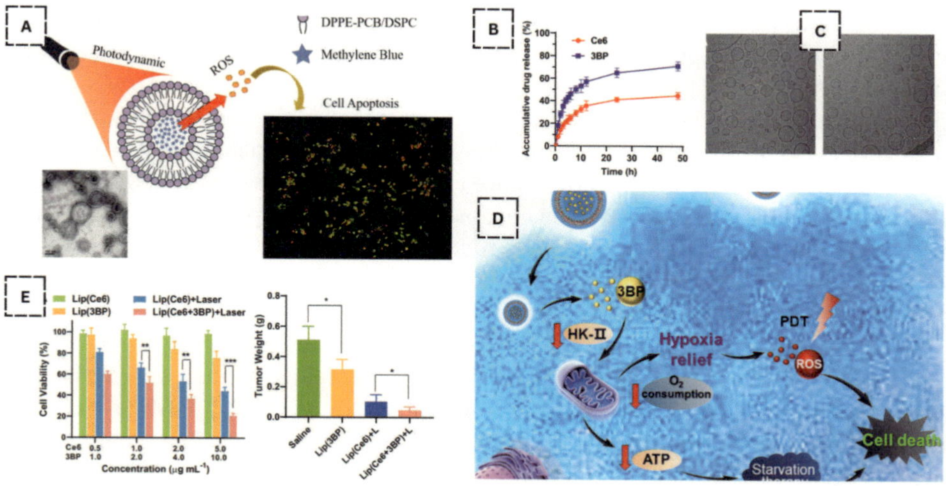

Figure 4.
(A) Scheme of the experiment with nanocapsules of LPs and methylene blue for PDT and ROS generation in breast cancer cells (4 T1 cells). (B) The accumulative release profile of Ce6 from liposomes nanocapsules, Lip(Ce6). (C) The average size of Lip(Ce6) measured by DLS. (D) Mechanism of synergistic photodynamic and starvation therapy with nanocapsules under laser irradiation. (E) In vitro cytotoxicity of Lip(Ce6) with light irradiation against HeLa cells measured by MTT assay, and weights of tumors after the tumor tissues were isolated from nude mice. From Wu et al. [58] and Li et al. [59], used under CC-BY 4.0.

range of 100 nm, represent a viable option for cancer therapy as their size is also suitable for improving permeability and biological stability [6]. Other authors also mention the co-encapsulation in LPs capsules with PSs such as porphyrin derivates [21, 64] and zinc phthalocyanine (ZnPc) [59, 65, 66] with excellent EE (85–90%), size distribution of 80–150 nm, and cytotoxicity effects under NIR irradiation.

Liposomes play a crucial role in enhancing the success of PDT and PTT by improving the solubility and bioavailability of photosensitizing drugs through nanoencapsulation. These lipid-based vesicles, formed by phospholipids, offer a range of advantages such as reduced toxicity, biodegradability, and biocompatibility. Their lipid composition allows for the encapsulation of hydrophilic and hydrophobic compounds, making them versatile drug carriers. Studies have shown successful encapsulation of several photosensitizers in liposomes, demonstrating high encapsulation efficiency and enhanced phototoxicity in cancer cells. Additionally, co-encapsulation of PSs like ICG and Ce6 in liposomes has also been investigated, resulting in stable and effective drug delivery systems with significant inhibition of cancer cell growth under near-infrared (NIR) irradiation.

2.4 Inorganic nanoparticles

Photosensitizers have been encapsulated in inorganic nanoparticles to enhance their efficacy in PDT and PTT. The precise characteristics of the agents and the intended use determine which nanoparticles are best for encapsulating photosensitizers with lower particle sizes. Calcium phosphate nanoparticles (CPNPs) [67, 68], mesoporous silica nanoparticles (MSN) [18], and hollow mesoporous-MnO_2 (H-MnO_2) nanoshells [69] have been investigated to enclose PSs and PTAs.

CPNPs are nontoxic, colloidally stable, nanoscale vehicles that can encapsulate ICG (ICG-CPNP nanocapsules) as a near-infrared imaging agent. ICG-CPNP shows an average size of 20 to 70 nm, with promise as a nanocarrier system for PDT and PTT treatments. These nanocapsules have proven to be successful in producing ROS generation and cytotoxicity to breast and pancreatic cancer cells, as well as capable of imaging tumors under 808 nm light irradiation [67, 68]. Similarly, MSN modified with trimethylammonium groups (MSN-TA) were used to enclose ICG (MSN-TA-ICG). MSN-TA-ICG demonstrated good stability and targeted delivery to malignant lesions, improving the effectiveness of PDT and PTT. Moreover, the *in vivo* biodistribution of MSN-TA-ICG revealed robust and stable fluorescence excitation and emission at 670 nm [18, 25]. Additionally, other studies focused on encapsulating PSs protoporphyrin (PpIX) and IR-820 NIR fluorophores into organically modified silica (ORMOSIL) doped with PEG. The PpIX/IR-820-doped OSMOSIL nanoparticles showed an average size of 42 nm with excellent PDT applications and control release [70].

Mesoporous silica nanoparticles (MSN) and hollow mesoporous H-MnO_2 nanoshells have extensively applied as carriers for PSs and as vehicles for encapsulating anticancer drugs. Zhou et al. demonstrated that MSN and H-MnO_2 can encapsulate anticancer drugs docetaxel and cisplatin (**Figure 5A**). These nanocapsules showed an average size of 80–100 nm and excellent applications in acid tumor microenvironments for the controlled release of drugs and excellent applications in PDT and PTT in metastasized oral squamous cell carcinoma [18, 69].

Inorganic nanoparticles have emerged as promising carriers for encapsulating PSs and PTAs to enhance the efficacy of PDT and PTT. The unique properties of nanoparticles such as CPNPs, MSN, and H-MnO_2 enable targeted delivery, controlled release, and enhanced therapeutic outcomes in cancer treatment. Encapsulation of

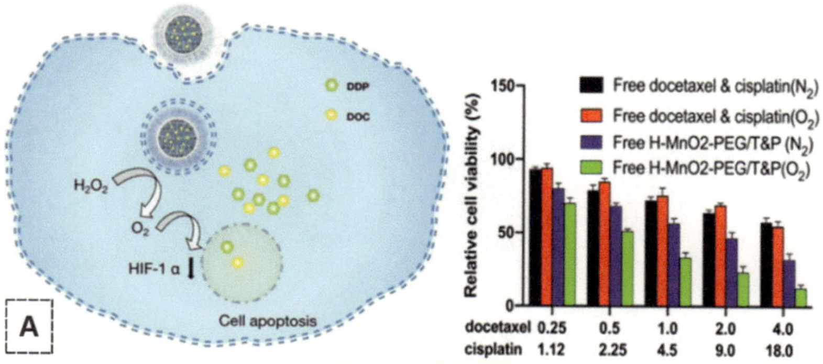

Figure 5.
H-MnO2-PEG/TP polymers nanoshells in pH-responsive drug delivery and cell viability in SCC7 and CAL27 tumor cells with and without H-MnO$_2$-PEG nanoshells in –N_2 or –O_2 atmospheres. From Zhou et al. [69], used under CC-BY 4.0.

PSs and anticancer drugs within these nanoparticles show great promise for improving treatment efficacy and precision in addressing various types of cancer. Overall, inorganic nanoparticles represent a valuable tool in advancing drug delivery strategies for PDT and PTT, contributing to the development of more effective and targeted cancer therapies.

2.5 Hydrogel nanocapsules

Hydrogel capsules are carrier systems at the micro and nanoscale that integrate the characteristics of hydrogels with capsules [23, 71]. These structures benefit drug delivery applications because of their special properties, such as compact size, high water content, and controlled release capabilities [72]. Polymeric nanogels and macromolecular micelles are sometimes referred to as hydrogel nanoparticles. Hydrogel capsules have emerged as a promising platform for encapsulating PSs and PTAs to PDT and PTT [73]. These nanocapsules are typically composed of hydrogel synthetic and natural polymers such as poly(methacrylic acid) (PMA) [74], poly(lactide-co-glycolide) (PLGA), poly(vinylcaprolactam) (PVCL) [75, 76], hyaluronan (HA), alginate [77], and chitosan [23], which are crosslinked to form a shell capable of encapsulating hydrophobic agent. The encapsulation of photosensitizers within these nanocapsules offers several advantages, including improved solubility, enhanced stability, controlled release, and targeted delivery to the site of interest [78].

The hydrogel nanocapsules are prepared using various methods such as sonication, nanoprecipitation, and ionic crosslinking, and their size ranges from 70 to 850 nm [78]. The encapsulation efficiency of the nanocapsules is high, ranging from 63.9 to 95%, and they exhibit sustained drug release profiles [15]. The nanocapsules have shown promising results in inhibiting tumor growth when applied in PDT.

Nan et al. demonstrated the properties of hydrogel nanocapsules as a nanocarrier system, developing hydrogel nanocapsules PAA-PNIPAm. PSS-PNIPAm were fabricated with poly(acrylic acid)-poly(N-isopropyl acrylamide) and encapsulated doxorubicin hydrochloride (DOX). The nanogels showed a high drug loading capacity, sustained release feature for water-solute drug molecules, and efficient ROS generation under 635 nm red light irradiation (**Figure 6A-D**) [74]. Similarly, Huang et al. encapsulated DOX into hydrogel capsules and a PSs zinc phthalocyanine (ZnPC). The hydrogel nanocapsules were developed through the polymerization of 3-caprolactone,

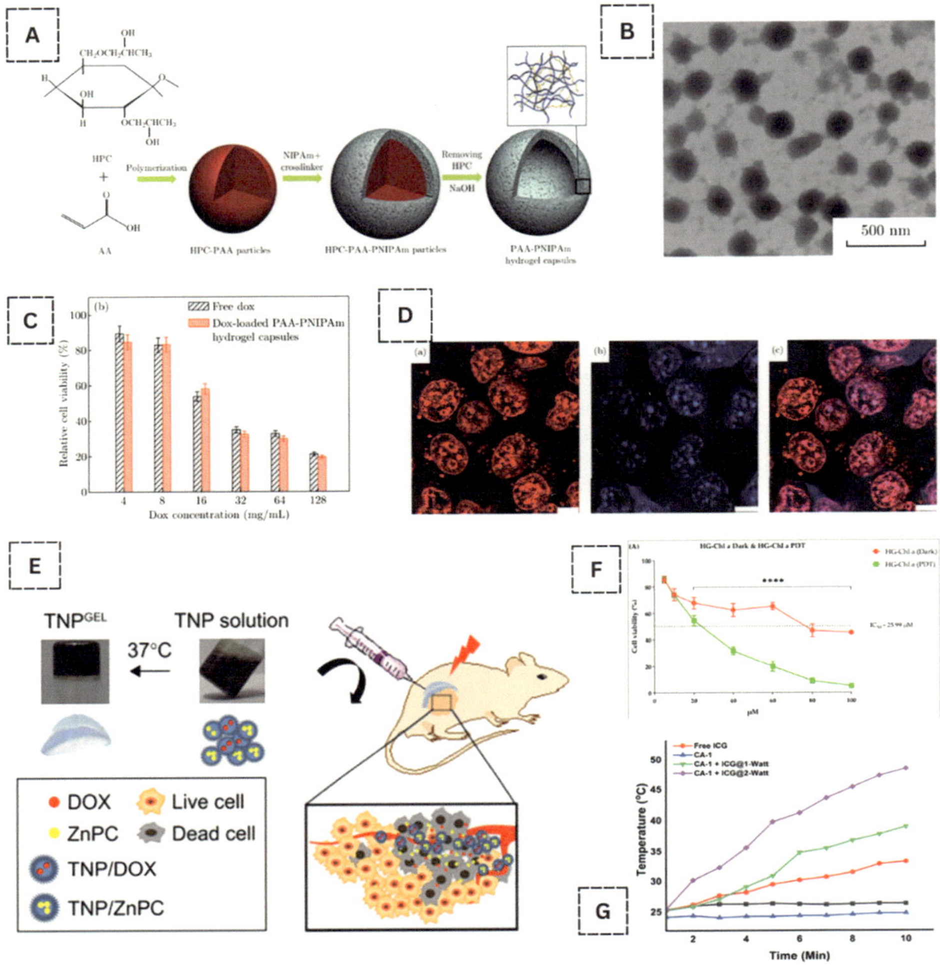

Figure 6.
(A) Schematic representation of the preparation of PAA-PNIPAm hydrogel capsules; (B) SEM image of PAA-PNIPAm hydrogel capsules; (C) dox-loaded PAA-PNIPAm hydrogel capsules, free dox against LoVo cell line at normal concentration; (D) LCSM images of 4 T1 cells incubated with dox-loaded PAA-PNIPAm hydrogel capsules for 2 h. From Nan et al. [74], used under creative commons CC-BY license. (E) Photo/chemo combination therapy via in situ formed thermal-sensitive polymer hydrogel TNP/ZnPC. From International Journal of Nanomedicine 2018 137623-7631 originally published by and used with permission from Dove Medical Press Ltd. (F) Cytotoxic activity of Cs-HGs containing Chl-A and Chl-A solutions against melanoma cells. From Araújo et al. [79], used under creative commons CC-BY license. (G) Photothermal effect of free PBS, free ICG, and formulated hydrogels (Alginate/CMC hydrogels) under different intensities of NIR irradiation. From Rizman et al. [77], used under CC-BY 4.0.

1, 4,8-trioxa[4.6]spiro-9-undecanone and PEG to form PCL-PTSUO-PEG and converted into thermal-responsive hydrogel nanocapsules through nanoprecipitation method obtained TNP/DOX and TNP/ZnPC (**Figure 6B**). The nanocapsules had an average size of 108 nm and EE of 95.2%, significantly improving aqueous solubility and *in vitro* drug release (**Figure 6E**). They also demonstrated high inhibition tumors (18.5%) in MCF-7 breast cancer cells under irradiation laser of 660 nm for 90 seconds and high-level ROS generation [57, 60, 80, 81]. Similarly, PVCL cross-linked with PEGDA was prepared through reversible addition fragmentation chain transfer (RAFT) polymerization to obtain nano-hydrogel through the precise control of polymer chain length and architecture, resulting in a polymer core-shell structure to encapsulate DOX. The anticancer efficiency of the drug-encapsulated structure was evaluated in breast cancer cell line MCF-7. The enhanced release rates are observed at acidic pH and elevated temperature, resulting in higher cellular uptakes and more cytotoxicity [82].

Hydrogel-based natural polymers have encapsulated PS and PTAs [23]. The potential of chitosan hydrogels to work as efficient carriers for delivering agents in PDT and PTT applications is being explored [19, 23, 83]. In a study, capsaicinoids-loaded nanocapsules were integrated into chitosan hydrogels. The hydrogel nanocapsules had a mean size between 70 and 100 nm and demonstrated excellent properties, great tissue adhesion, and release of nanoencapsulated drugs [19]. In a similar vein, Araújo et al. encapsulated PSs chlorophyll A (Chl-A) into chitosan hydrogels (HGs-CS-P407-Chl-A) for PDT against the murine melanoma cell line B16-F10. The HGs-CS-P407-Chl-A nanocapsules exhibited a thermosensitive system for topical applications and controlled release of Chl-A, thereby enhancing the efficacy of PDT against melanoma (**Figure 5F**) [41, 79, 84].

In contrast, the use of HA has been studied. Yi et al. demonstrated the permeability and retention of DOX into HA-NCs' stimuli-responsive hyaluronan hydrogel nanocapsules. The HA-NCs system exhibited great ability to release the encapsulated tumor cells, biocompatibility, and efficient cancer cell growth inhibition activity [85]. Similarly, alginate hydrogels have also been employed to demonstrate the photothermal impact of encapsulated ICG. The temperature of the ICG-loaded hydrogels rose under NIR irradiation, effectively eliminating tumor cells through photothermal means in HEK-293 cancer cells (**Figure 5G**) [77].

Overall, nanocapsules of hydrogel represent a versatile and promising approach for encapsulating and delivering agents for PDT and PTT, offering enhanced therapeutic outcomes and potential for clinical translation in the treatment of various cancers and other diseases. Moreover, hydrogel nanocapsules represent a promising approach in drug delivery for PDT and PTT due to their unique properties such as compact size, high water content, and controlled release capabilities. These nanocapsules, composed of synthetic and natural polymers such as PMA, PLGA, HA, alginate, and chitosan, offer advantages like improved solubility, stability, controlled release, and targeted delivery of PSs and photothermal agents PTAs. Various methods such as sonication, nanoprecipitation, and ionic crosslinking are employed to prepare hydrogel nanocapsules, with sizes ranging from 70 to 350 nm and high encapsulation efficiencies ranging from 63.9–95%. These nanocapsules demonstrate sustained drug release profiles and have shown promising results in inhibiting tumor growth when utilized in PDT.

3. Conclusions

- Nanocarrier systems, such as polymers, liposomes, hydrogels, inorganic particles, metal-organic frameworks, and covalent organic frameworks, offer unique advantages for enhancing the efficacy and safety of photodynamic therapy (PDT) and drug delivery.

- Polymers like poly(methacrylic acid) (PMA), poly(lactide-co-glycolide) (PLGA), and poly(vinylcaprolactam) (PVCL) have been utilized to form nanocapsules and hydrogels for encapsulation of photosensitizers, providing high encapsulation efficiency, sustained drug release, and enhanced cytotoxicity against cancer cells. However, further research is needed to optimize the particle size of nanocapsules for specific applications, aiming for enhanced cellular uptake and therapeutic efficacy. Exploring novel polymers and polymer combinations to enhance encapsulation efficiency and optimize drug release profiles is crucial for improving treatment outcomes. Research efforts are focused on investigating the potential synergistic effects of combining various polymers or incorporating targeting ligands to enhance specificity toward cancer cells.

- Liposomes, with their biocompatibility, high drug loading capacity, and ability to target specific tissues, have shown potential for improving the solubility and stability of photosensitizers and enhancing their accumulation at the target site. Continuing research aims to enhance the encapsulation efficiency and controlled release of photosensitizers through optimized liposomal formulations. Additionally, novel liposomal compositions and surface modifications are being explored to improve specificity and selectivity towards cancer cells.

- Inorganic particles, metal-organic frameworks, and covalent organic frameworks have also been explored to encapsulate photosensitizers, offering unique advantages such as tunable porosity, high surface area, and the ability to encapsulate hydrophobic and hydrophilic drugs. These carriers have shown potential for improving the stability and bioavailability of photosensitizers, as well as enabling controlled drug release, making them promising candidates for PDT and drug delivery applications. Future research and development in this area hold significant potential for advancing precision oncology and personalized cancer therapy.

Acknowledgements

Our heartfelt appreciation goes out to everyone who helped us finish this chapter. We express our gratitude to the scientists, researchers, and academics whose contributions and wisdom have greatly influenced the content of this publication. We would also like to thank the institutions and groups that gave the research and writing processes the tools and assistance they needed. Additionally, we thank our peers and colleagues in the field for their crucial advice and support, whose insights and experience have substantially improved the caliber of our work. Lastly, we thank the publishers and editors for their help and cooperation in making this chapter a reality.

Notes/thanks/other declarations

The chapter book on encapsulation of photodynamic and photothermal therapy agents provides a comprehensive overview of various carrier systems that offer unique advantages for enhancing the efficacy and safety of photodynamic therapy (PDT) and drug delivery. The chapter highlights the potential of polymers, liposomes, hydrogels, inorganic particles, metal-organic frameworks, and covalent organic frameworks for encapsulating photosensitizers, offering advantages such as high encapsulation efficiency, sustained drug release, enhanced cytotoxicity against cancer cells, biocompatibility, reduced toxicity, prolonged drug release, stimuli-responsive drug release, prolonged circulation time, and the ability to create a localized drug depot at the target site. The chapter also acknowledges the researchers, scientists, and scholars whose work and insights have been instrumental in shaping the content of this manuscript, as well as the institutions and organizations that have provided the necessary resources and support for the research and writing process. Finally, the chapter thanks the editors and publishers for their collaboration and assistance in bringing this chapter to fruition.

Appendices and nomenclature

ADG	andrographolide
AO	acridine orange
Ce6	chlorin e6
Chl-A	chlorophyll A
ClAlPc	chloroaluminum phthalocyanine
COF	covalent organic framework
CPDs	carbon-based polymer dots
CPNPs	calcium phosphate nanoparticles
Cur	curcumin
CXB	celecoxib
DCX	docetaxel
DOX	doxorubicin
EE	encapsulation efficiency
EPR	permeation retention effect
ICD	immunogenic cell death
ICG	indocyanine green
HA	hyaluronan
LPs	liposomes
MB	methylene blue
MOF	metal organic framework
MSN	mesoporous silica
NIR	near-infrared
NLCs	nanostructured lipid carriers
ORMOSIL	organically modified Silica
PCL	poly-E-caprolactone
PDT	photodynamic therapy
PEG	polyethylene glycol
PHBV	poly hydroxybutyrate-co-hydroxyvalerate
PLGA	poly(lactide-coglycolide)
PMA	poly(methacrylic acid)
PpIX	protoporphyrin IX
PS	photosensitizer
PTA	photothermal therapy
PVCL	poly(vinylcaprolactam)
RAFT	reversible addition fragmentation chain transfer
ROS	reactive oxygen species
SCC	squamous cell carcinoma
ZnPc	zinc phthalocyanine

Author details

María Paulina Romero[1]* and Myriam Alexandra González[2]

1 Escuela Politécnica Nacional (EPN), Quito, Ecuador

2 Laboratorio de Nuevos Materiales (LANUM), Escuela Politécnica Nacional (EPN), Quito, Ecuador

*Address all correspondence to: maria.romerom@epn.edu.ec

References

[1] Zeitouni NC, Oseroff AR, Shieh S. Photodynamic therapy for nonmelanoma skin cancers: Current review and update. In: Molecular Immunology. Elsevier Ltd; 2003;**39**:1133-1136

[2] Ge X, Liu J, Shi Z, et al. Inhibition of MAPK signaling pathways enhances cell death induced by 5-aminolevulinic acid-photodynamic therapy in skin squamous carcinoma cells. European Journal of Dermatology. 2016;**26**:164-172

[3] Li X, Chen L, Huang M, et al. Innovative strategies for photodynamic therapy against hypoxic tumor. Asian Journal of Pharmaceutical Sciences. 2023;**18**:1-28. DOI: 10.1016/j.ajps.2023.100775

[4] Salis F, Descalzo AB, Orellana G. Novel photosensitizing nanoparticles for PDT and biosensing applications. Journal of Photochemistry and Photobiology. 2021;**8**:100075

[5] Prathyusha EAP, Ahmed H, et al. Investigation of ROS generating capacity of curcumin-loaded liposomes and its in vitro cytotoxicity on MCF-7 cell lines using photodynamic therapy. Photodiagnosis and Photodynamic Therapy. 2022;**40**:103091

[6] Buddhiraju HS, Balaraman J, Dehariya D, et al. PLGA nanoparticle loaded with antioxidants and photosensitizer for ROS shock mediated phototherapy of triple negative breast cancer. Biomedical Materials (Bristol). 2023;**18**:3-12. DOI: 10.1088/1748-605X/acf5b9

[7] Cui Q, Yuan H, Bao X, et al. Synergistic photodynamic and photothermal antibacterial therapy based on a conjugated polymer nanoparticle-doped hydrogel. ACS Applied Bio Materials. 2020;**3**:4436-4443

[8] Li Y, Qi H, Geng Y, et al. Research progress of organic photothermal agents' delivery and synergistic therapy systems. Colloids and Surfaces. B, Biointerfaces. 2024;**234**:113743

[9] Oh Y, Jin JO, Oh J. Photothermal-triggered control of sub-cellular drug accumulation using doxorubicin-loaded single-walled carbon nanotubes for the effective killing of human breast cancer cells. Nanotechnology. 2017;**28**:1-30. DOI: 10.1088/1361-6528/aa5d7d

[10] Cai W, Gao H, Chu C, et al. Engineering phototheranostic nanoscale metal-organic frameworks for multimodal imaging-guided cancer therapy. ACS Applied Materials & Interfaces. 2017;**9**:2040-2051

[11] Zhou Z, Zhang L, Zhang Z, et al. Advances in photosensitizer-related design for photodynamic therapy. Asian Journal of Pharmaceutical Sciences. 2021;**16**:668-686

[12] Henn JG, Aguirre TAS, Nugent M, et al. Cancer nanomedicine: Recent developments in drug delivery systems and strategies to overcome eventual barriers to achieve a better outcome. Journal of Drug Delivery Science and Technology. 2024;**91**:105254

[13] Derycke ASL, de Witte PAM. Liposomes for photodynamic therapy. Advanced Drug Delivery Reviews. 2004;**56**:17-30

[14] Frank LA, Contri RV, Beck RCR, et al. Improving drug biological effects by encapsulation into polymeric nanocapsules. Wiley Interdisciplinary

Reviews. Nanomedicine and Nanobiotechnology. 2015;7:623-639

[15] Wang X, Yuan Z, Tao A, et al. Hydrogel-based patient-friendly photodynamic therapy of oral potentially malignant disorders. Biomaterials. 2022;**281**:121377

[16] Ding M, Liu W, Gref R. Nanoscale MOFs: From synthesis to drug delivery and theranostics applications. Advanced Drug Delivery Reviews. 2022;**190**:114496

[17] Ma S, Zhou Z, Zhang Y, et al. Review of irradiation treatments on MOFs and COFs: Synthesis, modification, and application. Separation and Purification Technology. 2024;**339**:126636

[18] Lee CH, Cheng SH, Wang YJ, et al. Near-infrared mesoporous silica nanoparticles for optical imaging: Characterization and in vivo biodistribution. Advanced Functional Materials. 2009;**19**:215-222

[19] Contri RV, Soares RMD, Pohlmann AR, et al. Structural analysis of chitosan hydrogels containing polymeric nanocapsules. Materials Science and Engineering C. 2014;**42**:234-242

[20] Bukreeva T, Barachevsky V, Venidiktova O, et al. Recent development of photochromic polymer capsules for smart materials. Materials Today Communications. 2024;**38**:107769

[21] Nawalany K, Rusin A, Kępczyński M, et al. Comparison of photodynamic efficacy of tetraarylporphyrin pegylated or encapsulated in liposomes: In vitro studies. Journal of Photochemistry and Photobiology. B. 2009;**97**:8-17

[22] Zelikin AN, Breheney K, Robert R, et al. Cytotoxicity and internalization of polymer hydrogel capsules by mammalian cells. Biomacromolecules. 2010;**11**:2123-2129

[23] Pellá MCG, Lima-Tenório MK, Tenório-Neto ET, et al. Chitosan-based hydrogels: From preparation to biomedical applications. Carbohydrate Polymers. 2018;**196**:233-245

[24] Tao C, Yu N, Ren Q, et al. Dressing and undressing MOF nanophotosensitizers to manipulate phototoxicity for precise therapy of tumors. Acta Biomaterialia. Epub ahead of print 2024;**17**:444-455. 2024. DOI: 10.1016/j.actbio.2024.01.042

[25] Huang C, Zhang Z, Guo Q, et al. A dual-model imaging theragnostic system based on mesoporous silica nanoparticles for enhanced cancer phototherapy. Advanced Healthcare Materials. 2019;**8**:1840-1900. DOI: 10.1002/adhm.201900840

[26] El-Daly SM, Gamal-Eldeen AM, Abo-Zeid MAM, et al. Photodynamic therapeutic activity of indocyanine green entrapped in polymeric nanoparticles. Photodiagnosis and Photodynamic Therapy. 2013;**10**:173-185

[27] Moharramnejad M, Malekshah RE, Salariyeh Z, et al. The synthetic strategies of COFs, for drug delivery, photo/sono-dynamic, photo/microwave thermal and combined therapy. Inorganic Chemistry Communications. 2023;**153**:110888

[28] Singla P, Parokie G, Garg S, et al. Enhancing encapsulation of hydrophobic phyto-drugs naringenin and baicalein in polymeric nano-micelles. Journal of Drug Delivery Science and Technology. 2023;**83**:104403

[29] Pramual S, Lirdprapamongkol K, Svasti J, et al. Polymer-lipid-PEG hybrid nanoparticles as photosensitizer carrier

for photodynamic therapy. Journal of Photochemistry and Photobiology. B. 2017;**173**:12-22

[30] Hou M, Chen W, Zhao J, et al. Facile synthesis and in vivo bioimaging applications of porphyrin derivative-encapsulated polymer nanoparticles. Chinese Chemical Letters. 2022;**33**:4101-4106

[31] Sooksai N, Treesuppharat W, Theeramunkong S, et al. Andrographolide-loaded nanoemulsion and its activity against non-melanoma skin cancer cells. In: Key Engineering Materials. Trans Tech Publications Ltd; 2019;**819**:139-144

[32] De Paula CS, Tedesco AC, Primo FL, et al. Chloroaluminium phthalocyanine polymeric nanoparticles as photosensitisers: Photophysical and physicochemical characterisation, release and phototoxicity in vitro. European Journal of Pharmaceutical Sciences. 2013;**49**:371-381

[33] Almeida EDP, Dipieri LV, Rossetti FC, et al. Skin permeation, biocompatibility and antitumor effect of chloroaluminum phthalocyanine associated to oleic acid in lipid nanoparticles. Photodiagnosis and Photodynamic Therapy. 2018;**24**:262-273

[34] Gao ZM, Xu X, Ruan LW. Enrichment and characterization of an anaerobic cellulolytic microbial consortium SQD-1.1 from mangrove soil. Applied Microbiology and Biotechnology. 2014;**98**:465-474

[35] Bazylińska U, Kulbacka J, Chodaczek G. Nanoemulsion structural design in co-encapsulation of hybrid multifunctional agents: Influence of the smart PLGA polymers on the nanosystem-enhanced delivery and electro-photodynamic treatment. Pharmaceutics. 2019;**11**:1-15. DOI: 10.3390/pharmaceutics11080405

[36] Zhang J, Li Y, Gao W, et al. Andrographolide-loaded PLGA-PEG-PLGA micelles to improve its bioavailability and anticancer efficacy. Expert Opinion on Drug Delivery. 2014;**11**:1367-1380

[37] Soares, Lima T, Silva de Oliveira MS, AVF R, et al. Nanoencapsulation of methylene-blue for enhanced skin cancer cell phototoxicity and cutaneous penetration in association with sonophoresis. Pharmaceutics. 2023;**15**:1-15. DOI: 10.3390/pharmaceutics15051371

[38] Daşkın D, Erdoğar N, İskit AB, et al. Oral docetaxel delivery with cationic polymeric core-shell nanocapsules: In vitro and in vivo evaluation. Journal of Drug Delivery Science and Technology. 2023;**80**:104163

[39] Doshi M, Copik A, Gesquiere AJ. Development and characterization of conducting polymer nanoparticles for photodynamic therapy in vitro. Photodiagnosis and Photodynamic Therapy. 2015;**12**:476-489

[40] Zhang X, Hou H, Wan J, et al. Inhibiting COX-2/PGE2 pathway with biodegradable NIR-II fluorescent polymeric nanoparticles for enhanced photodynamic immunotherapy. Nano Today. 2023;**48**:101759

[41] Denise, Prado Almeida E, André Santos Silva L, Rodolfo Souza de G, et al. Chitosan-functionalized nanostructured lipid carriers containing chloroaluminum 1 phthalocyanine for photodynamic therapy of skin cancer 2. European Journal of Pharmaceutics and Biopharmaceutics. 2022;**179**:221-231

[42] Potara M, Nagy-Simon T, Focsan M, et al. Folate-targeted

Pluronic-chitosan nanocapsules loaded with IR780 for near-infrared fluorescence imaging and photothermal-photodynamic therapy of ovarian cancer. Colloids and Surfaces. B, Biointerfaces. 2021;**203**:111755

[43] Lee SJ, Park K, Oh Y-K, et al. Tumor specificity and therapeutic efficacy of photosensitizer-encapsulated glycol chitosan-based nanoparticles in tumor-bearing mice. Biomaterials. 2009;**30**:2929-2939

[44] Jin Y, Yang S, Li F, et al. Fabrication of hordein/chitosan nanoparticles for encapsulation of curcumin. Journal of Food Engineering. 2024;**371**:111994

[45] Gao D, Gao Y, Shen J, et al. Modified nanoscale metal organic framework-based nanoplatforms in photodynamic therapy and further applications. Photodiagnosis and Photodynamic Therapy. 2020;**32**:102026

[46] Yan P, Fan J, Ji Y, et al. Photodynamic therapy strategy based on nanoscale metal-organic frameworks. Next Materials. 2024;**2**:100111

[47] Zou B, Xiong Z, He L, et al. Reversing breast cancer bone metastasis by metal organic framework-capped nanotherapeutics via suppressing osteoclastogenesis. Biomaterials. 2022;**285**:121549

[48] Shao Y, Liu B, Di Z, et al. Engineering of upconverted metal-organic frameworks for near-infrared light-triggered combinational photodynamic/chemo-/immunotherapy against hypoxic Tumors. Journal of the American Chemical Society. 2020;**142**:3939-3946

[49] Li W-Y, Kan J-L, Wan J-J, et al. A reactive oxygen species-responsive covalent organic framework for tumor combination therapy††electronic supplementary information (ESI) available. Chemical Communications. 2023;**59**:5423-5426. DOI: 10.1039/d3cc01136d

[50] Thankachan D, Anbazhagan R, Tsai HC, et al. Enhanced tumor targeting with near-infrared light-activated indocyanine green encapsulated in covalent organic framework for combined photodynamic therapy (PDT) and photothermal therapy (PTT). Dyes and Pigments. 2024;**221**:111812

[51] Ge L, Tang Y, Wang C, et al. A light-activatable theranostic combination for ratiometric hypoxia imaging and oxygen-deprived drug activity enhancement. Nature Communications. 2024;**15**:1-3. DOI: 10.1038/s41467-023-44429-y

[52] Li S, Chen Z, Tan L, et al. MOF@COF nanocapsule for the enhanced microwave thermal-dynamic therapy and anti-angiogenesis of colorectal cancer. Biomaterials. 2022;**283**:121472

[53] Fukui Y, Ito J, Takeiri M, et al. Preparation of a robust and degradable nanocapsule by polymer deposition over the liposome for loading and release of cargos. Colloids and Surfaces A: Physicochemical and Engineering Aspects. 2023;**676**:132160

[54] Germain M, Paquereau L, Winterhalter M, et al. Liposomes: support à la formation de capsules plus stables à base de polyéléctrolytes réticulés ou de silice. Annales Pharmaceutiques Françaises. 2007;**65**:134-141

[55] Zheng T, Chen Y, Liu L, et al. Liposome nano-capsule for target brain delivery. Biophysical Journal. 2019;**116**:426a

[56] Shramova EI, Filimonova VP, Frolova AY, et al. HER2-specific

liposomes loaded with proteinaceous BRET pair as a promising tool for targeted self-excited photodynamic therapy. European Journal of Pharmaceutics and Biopharmaceutics. 2023;**193**:208-217

[57] Pivetta TP, Ferreira Q, Vieira T, et al. Liposomes encapsulating methylene blue and acridine orange: An approach for phototherapy of skin cancer. Colloids and Surfaces. B, Biointerfaces. 2022;**220**:112901

[58] Wu PT, Lin CL, Lin CW, et al. Methylene-blue-encapsulated liposomes as photodynamic therapy nano agents for breast cancer cells. Nanomaterials. 2019;**9**:2-12. DOI: 10.3390/nano9010014

[59] Li X, Man J, Hu H, et al. Oxygen-economizing liposomes for synergistic photodynamic and starvation therapy. Colloid and Interface Science Communications. 2022;**47**:100598

[60] Swami Vetha BS, Oh P-S, Kim SH, et al. Curcuminoids encapsulated liposome nanoparticles as a blue light emitting diode induced photodynamic therapeutic system for cancer treatment. Journal of Photochemistry and Photobiology. B. 2020;**205**:1-22

[61] Shemesh CS, Hardy CW, Yu DS, et al. Indocyanine green loaded liposome nanocarriers for photodynamic therapy using human triple negative breast cancer cells. Photodiagnosis and Photodynamic Therapy. 2014;**11**:193-203

[62] Liu H, Wang Q, Guo J, et al. Prodrug-based strategy with a two-in-one liposome for Cerenkov-induced photodynamic therapy and chemotherapy. Journal of Controlled Release. 2023;**364**:206-215

[63] Szurko A, Rams-Baron M, Montforts F-P, et al. Photodynamic performance of amphiphilic chlorin e6 derivatives with appropriate properties: A comparison between different-type liposomes as delivery systems. Photodiagnosis and Photodynamic Therapy. 2020;**30**:101799

[64] Temizel E, Sagir T, Ayan E, et al. Delivery of lipophilic porphyrin by liposome vehicles: Preparation and photodynamic therapy activity against cancer cell lines. Photodiagnosis and Photodynamic Therapy. 2014;**11**:537-545

[65] Feuser PE, Cordeiro AP, de Bem SG, et al. Co-encapsulation of sodium diethyldithiocarbamate (DETC) and zinc phthalocyanine (ZnPc) in liposomes promotes increases phototoxic activity against (MDA-MB 231) human breast cancer cells. Colloids and Surfaces. B, Biointerfaces. 2021;**197**:111434

[66] Ceron Jayme C, Rezende NS, Fernandes D, et al. Target selectivity of cholesterol-phosphatidylcholine liposome loaded with phthalocyanine for breast cancer diagnosis and treatment by photodynamic therapy. Photodiagnosis and Photodynamic Therapy. 2022;**39**:102992

[67] Barth BM, Sharma R, Altinoğlu EI, et al. Bioconjugation of calcium phosphosilicate composite nanoparticles for selective targeting of human breast and pancreatic cancers in vivo. ACS Nano. 2010;**4**:1279-1287

[68] Altinoğlu EI, Russin TJ, Kaiser JM, et al. Near-infrared emitting fluorophore-doped calcium phosphate nanoparticles for in vivo imaging of human breast cancer. ACS Nano. 2008;**2**:2075-2084

[69] Zhang Z, Yuan LS, Chao ZT, et al. Overcoming chemotherapy resistance using pH-sensitive hollow MnO_2 nanoshells that target the hypoxic tumor microenvironment of metastasized oral

squamous cell carcinoma. Journal of Nanobiotechnology. 2021;**19**:1234-1312. DOI: 10.1186/s12951-021-00901-9

[70] Liang S. Protoporphyrin IX and IR-820 fluorophore–encapsulated organically modified silica nanoparticles. In: Bethesda, editor. Molecular Imaging and Contrast Agent Database. New York: National Center for Biotechnology Information; 2012;**12**:229-235

[71] Pérez-Luna VH, González-Reynoso O. Encapsulation of biological agents in hydrogels for therapeutic applications. Gels. 2018;**4**: 2-30. DOI: 10.3390/gels4030061

[72] Glass S, Kühnert M, Lippmann N, et al. Photosensitizer-loaded hydrogels for photodynamic inactivation of multiresistant bacteria in wounds. RSC Advances. 2021;**11**:7600-7609

[73] Saboktakin MR, Tabatabaie RM, Ostovarazar P, et al. Synthesis and characterization of modified starch hydrogels for photodynamic treatment of cancer. International Journal of Biological Macromolecules. 2012;**51**:544-549

[74] Nan J, Chen Y, Li R, et al. Polymeric hydrogel nanocapsules: A thermo and pH dual-responsive carrier for sustained drug release. Nano-Micro Letters. 2014;**6**:1234-1255. DOI: 10.5101/nml140022a

[75] Frost MJ, Lyons SG, Geever LM. Synthesis and characterisation of novel poly(N-vinylcaprolactam) hydrogels, cross-linked with N-hydroxyethyl acrylamide (HEA). Materials Today. 2019;**10**:436-440

[76] Romero JF, Diaz-Barrios A, Gonzalez G. Biocompatible thermo-responsive N-vinylcaprolactam based hydrogels for controlled drug delivery systems. Bionatura. 2021;**6**:1712-1719

[77] Rizwan A, Ali I, Jo SH, et al. Facile fabrication of NIR-responsive alginate/CMC hydrogels derived through IEDDA click chemistry for photothermal-photodynamic anti-tumor therapy. Gels. 2023;**9**:2-19. DOI: 10.3390/gels9120961

[78] Merino S, Martín C, Kostarelos K, et al. Nanocomposite hydrogels: 3D polymer-nanoparticle synergies for on-demand drug delivery. ACS Nano. 2015;**9**:4686-4697

[79] Araújo JL, da Silva PB, Fonseca-Santos B, et al. Photodynamic therapy directed to melanoma skin cancer by thermosensitive hydrogel containing chlorophyll A. Pharmaceuticals. 2023;**16**:2-15. DOI: 10.3390/ph16121659

[80] Hwang YM, Pan CT, Lin YM, et al. Preparation of biodegradable polycaprolactone microcarriers with doxorubicin hydrochloride by ultrasonic-assisted emulsification technology. Sensors and Materials. 2019;**31**:301-310

[81] Huang Z, Xiao H, Lu X, et al. Enhanced photo/chemo combination efficiency against bladder tumor by encapsulation of DOX and ZnPC into in situ-formed thermosensitive polymer hydrogel. International Journal of Nanomedicine. 2018;**13**:7623-7631

[82] Farjadian F, Rezaeifard S, Naeimi M, et al. Temperature and pH-responsive nano-hydrogel drug delivery system based on lysine-modified poly (vinylcaprolactam). International Journal of Nanomedicine. 2019;**14**:6901-6915

[83] Bugnicourt L, Ladavière C. A close collaboration of chitosan with lipid colloidal carriers for drug delivery applications. Journal of Controlled Release. 2017;**256**:121-140

[84] Mello VC, Araújo VHS, de Paiva KLR, et al. Development of new natural lipid-based nanoparticles loaded with aluminum-phthalocyanine for photodynamic therapy against melanoma. Nanomaterials. 2022;**12**:2-25. DOI: 10.3390/nano12203547

[85] Yi Q, Ma J, Kang K, et al. Dual cellular stimuli-responsive hydrogel nanocapsules for delivery of anticancer drugs. Journal of Materials Chemistry B. 2016;**4**:4922-4933

Chapter 3

Pectin-Based Encapsulation Systems for Bioactive Components

O.K. Sasina Sai, Usha K. Aravind and Charuvila T. Aravindakumar

Abstract

Pectin is a soluble dietary fiber with several health benefits, such as antibacterial, antioxidant, gastrointestinal-protective, and anticancer properties. Pectin is becoming an important class of materials owing to their inherent structural and functional properties such as biodegradability, binding potential, self-assembly, high nutritional value, gelling properties, non-toxicity, and good biocompatibility. Pectin is highly beneficial in microencapsulation since it allows for better control over the toxicity of the active substances and ensures the safety of the customer. Pectin offers a safe route for drug delivery due to its well-designed molecular architecture based on the changes in the biological process's fundamental mechanisms. The current arising insight into the chemical structure and associated health advantages of pectin opens new opportunities for the use of pectin in nutraceutical encapsulation and drug delivery. Pectin can be obtained from various plant sources at a lower cost. Thus, pectin is a promising biopolymer when designing materials that may achieve the highly desired dual objectives of being environmentally friendly and economically sustainable. This chapter emphasizes pectin-based nano and microencapsulation systems, their tailor-made functionalities, and their applications in the pharmaceutical and food industry.

Keywords: pectin, microencapsulation, drug delivery, biomaterials, nutraceuticals, nanoencapsulation

1. Introduction

The use of stable and active ingredients has taken center stage in both food and pharmaceutical industries as a result of the growing and expanding global knowledge of the value of a nutritious diet and its effect on preventing illness. Additionally, an increasing trend among consumers is their preference for foods made with natural ingredients. Many of such active ingredients lose their health benefits when it is added to a food matrix. One method for preserving the efficacy of food ingredients and promoting the creation of novel, nutritionally sound foods is the microencapsulation of these active food components [1]. The major active substances that encapsulate are essential oils, probiotics, minerals, vitamins, bioactive lipids, antioxidants, and enzymes [2]. Encapsulating substances to preserve flavor or prevent undesirable organoleptic qualities without sacrificing functionality is a commonly cited rationale.

Certain polymers have been used for encapsulation to preserve all of the functionality of the food components. Researchers are mostly focusing on microencapsulation methods using natural polymers. Because it is safer as well and consumers prefer products that use natural polymers. These polymers must satisfy regulatory standards, which can differ by nation and physicochemical requirements. Polymers used for microencapsulation generally need to have certain qualities; that is, the polymers used must be able to preserve all the properties of encapsulated active ingredients under elevated environmental conditions. However, when needed for the application, it should release the cargo, at right time and location [3]. Not to mention, it ought to be affordable and food-grade.

Proteins and carbohydrates are the main kinds of natural polymers that are most frequently used to encapsulate food. Each of these materials has benefits and drawbacks depending on the use. Certain components provide challenges when integrating them into the food matrix, while others may be expensive for numerous uses [4].

Pectin, a ubiquitous natural biomaterial, due to its unique advantages, has received particular attention in recent years when it comes to the encapsulation of food ingredients. Pectins are high molecular weight anionic heteropolysaccharides that, generally non-hazardous in nature, can form microencapsulation systems around active ingredients. As an encapsulating polymer, pectin offers many advantages. Pectin provides health advantages in addition to being useful for controlling and delivering active ingredients precisely to certain parts of the digestive tract. Pectin has the ability to stop gastrointestinal tract inflammation and may even stop some major illnesses, as this chapter will detail. These beneficial properties of pectin make it a promising polymer in the microencapsulation research area [5]. On the other hand, debates about the benefits and limitations of using pectin in food, as well as the kind of pectin that should be used in microencapsulation systems, are continuing. The potential of pectin to encapsulate particular food and pharmaceutical ingredients has been taken into consideration when writing the current book chapter.

2. Sources of pectin

Pectin is a polysaccharide present in plant cell walls that plays a role in intricate physiological processes like cell division and cell growth, which in turn affect the integrity and firmness of plant tissue. As a component of the cell wall, pectin is essential for giving plants structural support. The primary cell wall has a highly concentrated crosslinked network of cellulose and hemicelluloses, but pectin plays a crucial function in binding this network together [6]. According to structural research, pectin functions as a matrix into which the fibrils of cellulose and hemicellulose are inserted and fused to form a structural unit. Pectin is abundant in the middle lamella, and they are essential for the development of cell–cell adhesion in a variety of plant tissues [6].

Additionally, they are crucial to the defense systems against wounds and plant infections. Owing to their anionic characteristics, pectic polysaccharides are thought to have a role in regulating ion transport, wall porosity, and, consequently, the permeability of the walls for enzymes. They ascertain the water-holding capacity as well [7]. Fresh and processed food product's quality criteria are significantly influenced by the quantity and composition of pectin polymers found in fruits, vegetables, and other plant products.

It is possible to extract pectin from appropriate agricultural byproducts utilized in the food sector. Pectin is mostly produced commercially from apple pomace and citrus peels [8]. The pectin content in citrus peel is thought to be between 25 and 30 percent dry weight [9], compared to 10 to 15 percent in apple pomace. Pectin can also be extracted from other waste byproducts from industries, such as cocoa pod husks, grapefruit peels, pomegranate peels, passion fruit peels, mango peels, banana peels, kiwi fruit pomace, pistachio green hulls, tomato waste, sugar beet, sunflower head, pumpkin, watermelon rind, and potato pulp [10–23]. The need for the exploration of alternative pectin sources has arisen from the quick rise in demand for pectin owing to its extensive uses in various industries.

3. Physicochemical properties of pectin

Pectin is a complex heteropolysaccharide with significant intra- and intermolecular variability. Pectin is made up of segments of a long galacturonan chain as well as additional neutral sugars like rhamnose, xylose, galactose, and arabinose. Together with celluloses and hemicelluloses, it creates a matrix that supports cell structure.

According to a structural study, pectin is a polymer with a chain-like arrangement made up of between 100 and 1000 saccharide units; as a result, it lacks a clearly defined structure. Pectin is commonly represented as a heteropolysaccharide consisting of mainly three constituents: homogalacturonan (HG), rhamnogalacturonan I (RG-I), and rhamnogalacturonan II (RG-II). The other neutral sugar units including xylogalacturonan (XG), Arabinan, arabinogalactan I (AG-I), and arabinogalactan II (AG-II) may branch with the backbone structure. Around 60% of all the pectins in cell walls are homogalactalacturonan, a simple, non-substituted chain of polygalacturonic acid known as "smooth" chains. The structural description of Rhamnogalacturonan I (RG-I) is a lengthy backbone sequence of alternating α-(1 → 4) linked galacturonic acid (GalA) units and α-(1 → 2) linked D-rhamnose, consisting of spatially organized polymers. It may get O-acetylated at the GalA residues' O-2 or O-3 sites. Whereas rhamnogalacturonan II is made up of a homogalacturonan backbone switched with a wide range of complex side chains of glycan, comprising several types of neutral sugars, GalA, which is found in the (RG-II) backbone, might be methyl esterified at C-6 position. Xylogalacturonan (XG) is made up of the α-(1–4) linked D-GalA in the backbone linked by xylose in branches. The α-(1–4) glycosidic linkages are present throughout the homogalacturonan pectin chains. Depending on the pectin source, the amount of GalA residues in HG can range from 72 to 100%. Moreover, it was noted that HG might be O-acetylated at the O-2 and/or O-3 and/or methoxy-esterified at the C-6. HG can be degraded and de-esterified by both enzymatic and mechanical means. Galactan and arabinan polymeric side chains are substituted at the RG-I backbone's O-4 position. It has also been observed that Arabinogalactan I (AG-I) and Arabinogalactan II (AG-II) exist as polymeric side chains. The side chains, often known as "hairs," are thought to be crucial to the functioning of pectin. The solubility of the pectin may increase as a result of side chain loss [24].

Pectin molecules have a certain degree of methyl-esterification and acetylation on the galacturonic acid units. For the time of the esterification reaction of pectin, the hydrogen on the carboxylic acid group is changed to a methyl group (CH_3), transforming the R group from COOH to $COOCH_3$. The degree of esterification of D-galacturonic acid refers to the proportion of its carboxyl groups that have undergone methyl group esterification. The esterification of the galacturonic acid groups with methyl groups at C-6 and acetyl groups at O-2 and O-3 on the homogalacturonan is the basis for the specific structural arrangement of pectin [24].

The pectin charge is mainly determined by the presence and distribution of esterified and nonesterified galacturonic acids. Pectins are classified as high-methoxy pectin (HMP) or low-methoxy pectin (LMP) based on their degree of esterification (DE). The degree of esterification is less than 50% in LMP, whereas HMP pectin has DE values above 50% [25]. Besides, it was ascertained that the viscosity, solubility, and gelation characteristics of pectin are strongly correlated with structural characteristics [24].

4. Pectin as a functional biomaterial

Pectin is a weak acid, that contains carboxyl, hydroxyl, acetyl, and ester functional groups for chemical response. It has a pKa range of 2.9–5.4. The functional hydroxyl groups at C2 or those at C6 located in the equatorial site of pectin are chemically more reactive. Pectin's interactions with various small and macromolecules, as well as with each other, are determined by its structural and molecular properties. Pectin can help to remove toxic heavy metals such as mercury, lead, arsenic and cadmium through the mechanism of metal chelation. Pectin binds to these toxic metals and forms complex structures that are able to be removed from the body through excretion. Pectin with a lower degree of esterification is rich in free carboxyl groups, which is ideal for the chelation of cations or metals [26].

The pH of the stomach is 2 or below. In this stomach environment, the carboxyl groups in pectin become protonated. As a result, there is less pectin-cation binding and a greater cation release. In the small intestine (pH 6-7), pectin is highly charged due to the ionization of carboxyl groups at higher pH. This will promote the electrostatic interaction between pectin and cations [26]. In the colon or large intestine, there are microflora that can ferment pectin, thereby increasing the release of bound divalent cations. Thereby increasing the potential for cation absorption [27]. The high viscosities and high molecular weight of pectin can help to function as a natural barrier around the oxidative enzymes and can suppress the actions [28]. The carboxyl and hydroxyl functional groups of pectin can act as a natural donor of electrons for free radicals quenching and exert pectin's immunomodulatory and antioxidant potential [26].

Furthermore, it is possible that the high-viscosity pectin can form a thicker layer of mucous on the intestinal mucosa surface, thereby limiting the intestinal mucosa's ability to absorb ingested carbohydrates and decreasing glucose absorption [29]. Pectin can slow down the contact time between ingested foods and enzymes in the intestine by holding food components in the intestine and by providing a thick layer coating on food fragments, which reduces the amount of food that is absorbed [30, 31]. Due to their higher water-binding capacity, pectin can provide a feeling of fullness and is useful in treating overeating disorders [32]. Furthermore, pectin's capacity to extend stomach emptying and produce a longer-lasting sensation of "fuller" may significantly reduce the urge to eat [33, 34]. By slowing down the emptying of the stomach, reducing the diffusion of glucose, and reducing the absorption and storage of fat, pectin has been linked to weight loss [35–37]. Because pectin can form gels to swell in the aqueous surroundings of gastric fluids through self-association, this can stick to the stomach walls and bind to bile acids and the micelle components, such as cholesterol, free fatty acids, and monoglycerides, to decrease the absorption of fat in the stomach [38–40]. The high methoxy pectin can form a gel network with mucin, present in the intestinal epithelium via hydrogen bonds, and can give strength to the mucus layer and provide protection of intestinal epithelium against unwanted mucus penetrating materials. Meanwhile, low-methoxy pectin is capable of penetrating the mucus membrane to get into the intestinal epithelium and can excite intestinal mucosal epithelial cells to secrete the mucus. Thereby promote the protection of intestinal epithelium. Pectin can be degraded by pectinolytic enzymes produced by the gut microbiota in the colon. The decomposed pectin is used by intestinal microflora as prebiotics; through this, pectin can boost the growth and viability of probiotic strains from the gut and can contribute to the strengthening of the gastrointestinal immune system. Pectin can have an anti-adherence effect on pathogens and promote the adhesion of beneficial bacteria. This property can help to replace the pathogenic bacteria in the intestine [26]. In addition, pectin showed cholesterol-decreasing effects in humans [41–44].

Pectin may aid in combating different types of cancer in several ways. Pectin has been shown to enhance immune function, control blood sugar, support the growth and viability of probiotics, regulate oncogenes, cause cancer cells to undergo apoptosis, inhibit the growth and development of tumors, and have effects that are anti-inflammatory, antioxidant, and wound healing [45, 46]. Pectin and its derivatives have been shown in several preclinical studies to effectively suppress cancer cells including the colon [47], liver [48], prostate cancer [49], bladder cancer [50], breast [51], and lung [52]. Pectin has intrinsic antitumor potential because it can identify and bind to the galectin-3 receptors that are found on different types of cancer cells. This has led to the evolution of many pectin-based nano-formulations for the delivery of chemotherapeutic medications [26]. Numerous systemic illnesses, including cancer, may be made worse by gut inflammation [53]. Pectin can modify symbiotic flora, inflammatory-associated cytokines, chemokines, intercellular adhesion molecules, and the gastrointestinal (GI) immune system [54–56]. Pectin is thought to have anti-inflammatory and immunomodulatory properties through a microflora-independent mechanism such as inhibiting the deterioration of the colonic mucosa barrier [57], influencing nonimmune and immune cells and attaching themselves to pathogens or toxins [58], and through a microflora-dependent pathway by associating with stomach bacteria, enhancing their metabolic processes, healing gut microorganism disorders, and delivering anti-inflammatory properties [56].

Moreover, atherosclerosis, metabolic problems (obesity and diabetes mellitus), diseases of the liver and pancreas, and hypoglycemia are all treated and prevented with pectin. It has been established that pectin act as protective agents against enzymatic proteolysis. Pectin-stabilized polypeptide medications, therefore, remain intact in the stomach and small intestine and subsequently release drug molecules once pectin is broken down by the colon's microbiota [26]. One drawback of utilizing pectin by itself is that, due to its tendency to swell under physiological conditions, it releases drugs prematurely from the body [48]. Pectin mixtures with other polymers help to reduce this effect.

5. Microencapsulation

The technique known as "microencapsulation" involves the physicochemical or mechanical encapsulation of one material in another to create particles that are between nanometers and millimeters. The benefits of microencapsulation, including enhanced thermostability, protection of bioactive compounds, controlled release, preservation of volatiles, odor shelter, and improved texture/sense, have led to its strong recommendation in the food sector these days. A range of molecular interactions, such as van der Waals force of attraction, electrostatic force of attraction, and hydrogen bonding, are used to microencapsulate core materials in multilayers of wall materials. The wall materials are engineered to guide the core to its intended place while accommodating varying environmental conditions. Ultimately, as the micro-encapsulation system passes through the human digestive system, it should be able to dissolve in a gastric environment, and then the encapsulated core ingredients are delivered and absorbed in the intestine [1]. Pectin is a promising encapsulation material due to its many advantages, including its ability to stabilize emulsions, non-toxic nature, gelling, and binding properties [59].

5.1 Techniques used for the development of microencapsulation systems

The microencapsulation system using pectin can be created using different techniques. The major ones are discussed below.

5.1.1 Spray-drying technique

In the spray-drying method, the microencapsulation system is generated by atomizing the emulsion of the wall and core materials in a high-temperature environment. This process quickly hardens the droplet shell to encase the core material and evaporates moisture by heat transfer between the drying medium and droplets. This is the most popular embedding technology. This coating approach is cost-effective (30–50 times less expensive than freeze-drying), easy to use, capable of continuous production, and suitable for large-scale manufacturing. Better water solubility and low viscosity are necessary for wall materials to be suitable for the spray-drying process [1, 60].

5.1.2 Emulsification technique

Emulsification involves a chemical embedding process in which a mixture of dispersed phase, that is, core and wall materials, is combined with a significant amount of continuous phase, that is, vegetable oil containing the emulsifier. And form a stable emulsion, then microencapsulate with the help of a crosslinking agent. With the ease of use and excellent survival rate of encapsulated bioactive components, this technology allowed for a realistic preparation process; however, the enormous volume of vegetable oil required meant that manufacturing costs were typically very high [1].

5.1.3 Freeze-drying technique

The sublimation of ice into vapor using high vacuum conditions following a rapid freeze is known as freeze-drying. The process of ice sublimation eliminates heat and maintains a cool environment, hence maintaining the biological samples' activity, including proteins. Nevertheless, the microbial cell membrane's integrity may be compromised by the production of ice crystals and the higher osmotic pressure during the freeze-drying process. For this reason, hydrophilic material is typically added to the solution as a cryoprotectant. The freeze-drying technology is more often employed for high-value, heat-sensitive foods due to its high cost [1].

5.1.4 Coacervation technique

In the coacervation method, after suspending the core ingredient in the wall material solution, a different solvent or material is introduced to lessen the solvability of the wall material, which is then uniformly aggregated and occupied around the core substance to create microcapsules. Both the complicated condensation and the single condensation are part of the coacervation procedure. Two oppositely charged materials are employed as wall materials in the complicated coacervation, while the core ingredients are emulsified and distributed within the solution of wall materials. By interacting with the opposing charges, the two wall materials and the core substance combine to produce microcapsules by controlling the pH, concentration, and temperature of the system's aqueous solution. Complex coacervation is a popular method for encasing fat-soluble food materials. It has the following benefits: less damage to the quality of the core material, mild process conditions, antioxidation efficiency, controlled core release properties, and higher product encapsulation efficiency [1, 60].

5.1.5 Layer-by-layer self-assembly technique

In LbL assembly techniques, the multilayers spontaneously adhere to one another to create stable supramolecular structures or molecular aggregates with particular performances or functions resulting from noncovalent interactions such as hydrogen bonds, coordination bonds, and electrostatic attraction. Accurate nanoscale control over the capsule's shape, size, thickness, composition, and structure is achieved through the use of LbL assembly technology. As a result, LbL assembly technology presents itself as a viable technique for creating multilayer microcapsules that adapt to changing environmental conditions [1, 59].

5.1.6 Extrusion technique

Squeezing the colloid mixture and core material into the hardening bath as liquid droplets through using a needle-like tube under pressure is known as extrusion, and it is a physical embedding technique used to make microcapsules. The cost of this method is twice as high as that of the spray-drying method; however, the formed micropores have a small surface area, which allows them to successfully protect oil and oxygen from volatilization and improve product shelf-life. Pigment, vitamin, and volatile chemicals of all kinds, as well as other heat-sensitive components, are frequently embedded using the extrusion approach [5, 60].

5.1.7 Electrospray technique

Electrospray is a technique that breaks down the polymer fluid into small droplets using an electric field. This procedure involves a polymeric solution that is flowing out of a capillary nozzle, which is kept at high potential and subjected to an electric field. A jet forms when the electric field reaches a threshold value. The jet then deformed and dispersed due to the electric field, creating droplets. Once the solvent has evaporated, fine polymer particles are produced. Which is then collected using a metal collector to get the microcapsules. This eco-friendly method can be implemented without the need for extra reaction solvents. The uniform and nanoscale size of microcapsules created via electrostatic spraying has garnered increasing interest. The food and pharmaceutical industries are making significant use of the electrospray technique due to its immense potential for encapsulating volatile molecules, bioactive compounds, controlled-release additives, and functional foods [1, 59].

6. Pectin-based encapsulation systems and its applications in nutraceutical encapsulation

Nutraceuticals are dietary supplements that provide therapeutic or health benefits. These supplements contain bioactive nutrients in significant concentrations and dosages compared to the nutrients that our bodies ordinarily absorb from a regular diet. Dr. Stephen D. Feliz, who is regarded as the founder of nutraceuticals, first used the phrase "nutraceutical" in 1989. Thus, the term nutraceuticals came from nutrition and pharmaceuticals. In recent years, the nutritional value of several foods has been diminishing due to the use of hazardous chemicals, pesticides, and artificial fruit and vegetable ripening in agricultural practices. In the long term, a lack of vitamins, minerals, proteins, carbs, fats, and other nutrients will result in a variety of diseases due to insufficient daily intake. Nutraceuticals are a key component in closing the nutrition gap and achieving optimal nutrition [61]. Pectin is incredibly effective at increasing the stability and bioavailability of encapsulated nutraceuticals. The commonly utilized pectin-based encapsulation systems and their applications in nutraceutical encapsulation are covered in the section below.

6.1 Pectin-based hydrogels

Hydrogels are materials composed of crosslinked polymers in three dimensions that possess the capacity to absorb and hold vast amounts of water. According to studies [62, 63], hydrogels are used to enhance the water solubility and release characteristics as well as the stability of encapsulated compounds. They can be engineered to respond to stimuli by shrinking, swelling, as well as disintegrating in response to particular environmental factors including pH, enzyme activity, temperature, or ionic strength. The hydrogels' pores can encapsulate bioactive materials [64, 65]. Hydrogel particles have the capacity to enclose bioactive substances that are both hydrophilic and lipophilic [66]. Proteins and polysaccharides, such as pectin, are the widely utilized polymers to make hydrogels [67]. The biocompatible and biodegradable nature of pectin is a clear advantage for encapsulating bioactive components. However, pectin formulations have less mechanical strength, which may make them susceptible to low drug loading efficiency and premature drug release. The drawbacks of pectin can be alleviated by developing pectin-based formulations using physical (polymer blending) and chemical (graft copolymerization and amidation) modifying techniques.

By using the graft copolymerization technique, scientists created a pH-sensitive polyacrylamide grafted pectin hydrogel that excelled at pectin in terms of gelling and film-forming ability. They discovered that the formed hydrogels were biocompatible and showed pH sensitivity and increased cell viability of B-16 melanoma cells [68]. In another study, they altered pectin using the amidation reaction, employing methanol as the solvent. The ethanolamine chemically modified pectin, and the resultant product, was used to create hydrogel that was crosslinked with glutaraldehyde. A substantial increase in gelling and film-forming capabilities was observed upon the addition of amide groups to pectin. They found that amidated pectin has a good swelling property. Which helps the effective release of salicylic acid at colon pH for an extended period. The hydrogel created in this study was found to have good cell viability with B-16 melanoma cells [69].

Peng et al. employed bovine serum albumin (BSA) and citrus peel pectin to make nanohydrogel by using a self-assembly approach to encapsulate vitamin C. After 10 days of storage, they discovered that the nanohydrogel system maintained its stability at about 73.95% and that the efficiency of vitamin C encapsulation was approximately 65.31%. They also concluded that this hydrogel that self-assembles could be used as a possible delivery method to increase the stability and bioavailability of functional chemicals [70]. In a different study, curcumin was delivered orally using nanogels made of low-density lipoprotein and pectin [71, 72]. Consequently, the generated nanogels facilitated the controlled release of curcumin and showed a fine surface with a homogeneous distribution in size.

Various researchers have also employed a modification of pectin to create drug-delivery devices. Commercial low-methoxy pectin and charge-modified high-methoxy pectin were utilized to develop hydrogel beads to encapsulate the drug, indomethacin [73]. The researchers discovered that the efficiency of the encapsulating system was improved with the use of modified pectin, which helped control the release of indomethacin at various pH levels. In another study, encapsulated doxorubicin (an anticancer medication) in a hydrogel that used oxidized pectin. According to their findings, doxorubicin-releasing oxidized pectin hydrogels can stop the development of metastatic cancer in addition to slowing the advancement of primary cancer [74].

6.2 Pectin-based micro/nanoparticles

Polysaccharide-based micro/nanoparticles are classified as colloidal particles since they primarily contain polysaccharide molecules inside of them instead of water. The physicochemical features and functional performance of this type of encapsulation system, including its stability, encapsulation effectiveness, loading capacity, and release profile, can be modified by adjusting its size, shape, composition, and charge [67]. Pectin-based micro/nanoparticles take on more importance, due to their biodegradable nature and ease of digestion by the intestinal microbiota.

Yu et al. created composite microparticles using pectin, alginate, and chitosan for oral delivery of proteins. Bovine serum albumin (BSA) was used as a model protein. In this study, the microparticles were created by using tripolyphosphate crosslinking, ionic gelation of calcium ions, and electrostatic complexation of alginate and pectin. They proposed that target-specific protein delivery through oral administration could be facilitated by the microparticles due to its better pH sensitivity [75].

Jones and team examined how the charge density of polysaccharides affects the generation and characteristics of nanoparticles made by thermal treatment of β-lactoglobulin-pectin complexes. They proposed that the type of pectin utilized affects the stability and size of the particles; that is, high, methoxy pectin produces smaller and more pH stable particles than low-methoxy pectin. Then, the biopolymer nanoparticles seem to be mostly composed of aggregated protein molecules; however, at pH values where there is sufficient electrostatic attraction between protein and polysaccharide, they are most likely complexed with pectin. They concluded that the developed protein–pectin nanoparticles may utilized as encapsulation systems in food industries [76].

Pectin was investigated for the suppression of agglomeration of human serum albumin (HSA) nanoparticles loaded with ciprofloxacin [77]. By using the pH-coacervation approach, HSA-pectin nanoparticles loaded with ciprofloxacin were created, and their various physicochemical properties were assessed. The scientists proposed using pectin as a pharmaceutical additive to decrease the particle agglomeration in HSA nanoparticles. Verma et al. created pectin nanoparticles and evaluated their effectiveness as a drug delivery system for paclitaxel, an anticancer drug. The authors proposed that the amide group of L-asparagine and the carboxylate and hydroxyl ions of pectin were actively involved in the process that formed the nanoparticles. Over the course of a month, the drug's release pattern was biphasic, releasing 96% of the drug at pH 5.8 and 81% at pH 7.4 [78].

Resveratrol is a polyphenol derived from plants. It has garnered attention recently because of its potential health benefits, which include cardio and neuroprotection, hepatoprotection, anti-aging, antioxidant, anti-carcinogenic and anti-inflammatory properties [79]. Huang et al. encapsulated resveratrol using pectin–zein complex nanoparticles. This approach produced nanoparticles with a yield of 91.7% and a resveratrol content of about 10.2% (w/w). The nanoparticle loading efficiency of resveratrol was 77.9%. When compared to free resveratrol, the resulting encapsulated form of resveratrol exhibited increased antioxidant and anticancer effects. According to the study's findings, resveratrol-loaded biopolymer nanoparticles may be useful in health supplements and food and beverage products [79]. In another study, the resveratrol was encapsulated in gellan gum/pectin blend nanoparticles using the ionotropic gelation process. The permeability of the nanoparticles and their sustained release pattern was evaluated in the Caco-2 cell model and the mucus-secreting tripe co-culture model [80]. They found that particles can load high concentration of drug (over 80%) in an acidic media with a low pH in 2 h and sustained resveratrol release of up to 85% in 30 hours in a pH 6.8 media.

In another investigation, pectin nanoparticles were created utilizing food-grade cassava root material to coat the nutritional supplement β-carotene, with the goal of increasing the bioavailability of these lipophilic dietary nutraceuticals during gastro-intestinal (GI) delivery. The particles have a mean size of 21.3 nm, are mucoadhesive, and have enhanced antioxidant activity, bioavailability, and stability up to 90 days at 4°C [81]. Researchers developed caseinate/pectin-based phytosterol nanoparticles by combining emulsification, evaporation, and complex coacervation processes. These protein/polysaccharide composite nanoparticles demonstrated a significant phytosterol loading capacity (21%) and encapsulation efficiency (91%). The findings showed that the nanoparticles effectively encapsulated and protected the phytosterols while remaining stable throughout a 15-day storage period at 4°C and 25°C [82].

Curcumin-loaded nanoparticles with over 86% encapsulation effectiveness were created by a study [83] using zein-pectin complexation. According to their findings, the same method's nanoparticles can be useful for incorporating curcumin as a pharmaceutical product. In another study, high methoxy pectin and sodium caseinate were employed to create nanoparticles, and it was discovered that the electrostatic interaction between the sodium caseinate and high methoxy pectin-based nanoparticles was stable across a range of pH values [84].

Eugenol essential oil is a natural phenolic compound derived from cloves and is valued for its potent antibacterial and antioxidant properties [85]. According to a study, zein-pectin-sodium caseinate complex nanoparticles can encapsulate eugenol. As an amphiphilic protein, sodium caseinate forms the intermediate layer between the hydrophilic pectin polymer and the hydrophobic zein polymer. The highly hydrophobic zein protein is positioned in the center of the complex structure of this nanoparticle, while pectin is in the outermost layer due to its higher affinity toward water [85].

6.3 Pectin-based nanofibers

Nanofibers-based encapsulation systems are composed of fibers that have sizes between one and one hundred nanometers [86]. These materials' high surface area to volume ratio is beneficial for the sustained release of several active ingredients [87]. Electrospinning is one of the most utilized methods to make nanofibers due to its ease of use and efficiency [88]. Polysaccharide solution is pulled through a pump in the presence of a high electrical field during the electrospinning process. This creates a jet of fluid followed by solvent evaporation and solidification. Then, the resulting nanofibers deposit on the collector plate [89]. The crosslinking capacity of pectin-based formulations makes them suitable for the preparation of nanofibers. Additionally, due to their ability to mimic the structure of the natural extracellular matrix, electrospun pectin nanofibers offers unique advantages.

Tomasula et al. created ultrafine fibers using the electrospinning method. They used pullulan and pectin to create fibers. *Lactobacillus rhamnosus* GG, a probiotic, is used as a model bioactive component to demonstrate how the generated fibers may incorporate and preserve bioactive compounds. They showed that the generated fibrous mats have a potential for the food industry [90]. In another investigation, water-resistant pectin nanofibers were effectively produced using electrospun synthesis with different types of pectin (HMP and LMP). This study concluded that drug delivery could be facilitated by the use of manufactured nanofibers [91]. Folic acid was encapsulated in pectin-poly (ethylene oxide)-alginate electrospun nanofibers [92]. The study revealed that folic acid encapsulated in the pectin-alginate system shows high folic acid retention when compared to alginate alone encapsulation system. The recovery of folic acid encapsulated with nanofibers was about 100% when it was stored in the dark at pH 3 for nearly 41 days. On the other hand, the retention of unencapsulated folic acid was found to be insignificant. Furthermore, other studies also successfully created electrospun-coated nanofibers utilizing polyethylene oxide and pectin, with the conclusion that these pectin-based nanofibers have prospective uses in biomedicine [93, 94].

6.4 Pectin-based micro/nanocapsules

Nanocapsules are small particles with a core and shell structure. The protection and distribution of bioactive substances have been extensively employed with nanocapsules [95]. A polymer shell encircling a fluid center is their typical structure. Since less polymer is needed to generate the delivery systems with nanocapsules than with nanoparticles, this is an advantage [96]. The utilization of polysaccharide-shelled nanocapsules has been observed to mitigate immunological elimination and extend their circulation in the body [97].

A study investigated the encapsulation of *Bifidobacterium* probiotics in pectin-and alginate-based microcapsules using the emulsification/internal gelation process. When compared to its alginate counterpart, pectin, as the wall material used in microencapsulation, had a better encapsulation yield (89.5–96.9%). It also retained a higher probiotic viability after freeze-drying and in the simulated infant gastrointestinal digestion [98]. Osteopontin (OPN), an acidic glycoprotein, usually found in human and domestic animal milk. It has been related to early life developments, including immunological modulation, intestinal development, and nervous system development [99]. They found that, using OPN as the excipient for pectin-based microencapsulation systems further enhanced the probiotic viability even more in challenging circumstances. After intestinal digestion, non-microencapsulated *Bifidobacterium bifidum* R0071 and *Bifidobacterium breve* M-16 V showed markedly decreased viability. While all strains of *Bifidobacterium* in pectin and alginate-based microcapsules retained much higher viability (8–9 log CFU.mL^{-1}). Even though the intestinal phase of the microencapsulation showed a decline in probiotic viability compared to the stomach phase, encapsulation was still able to effectively shield *Bifidobacterium* strains throughout intestinal digestion [98]. In a different investigation, the *Lactobacillus acidophilus* LMG9433 encapsulated microcapsule using low-methoxy pectin developed and found that probiotics were significantly protected during intestinal digestion. They discovered that the microcapsule had higher probiotic viability (8 log CFU.mL^{-1}) in comparison to the free bacterial cell sample (5 log CFU.mL^{-1}) [100].

Whey protein–pectin nano spray-dried microcapsules were created by a study to encapsulate grape marc phenolics and investigate the impact on the stability and bioaccessibility of the polyphenols. Enhancing bioaccessibility, the new encapsulating technique preserved the antioxidant activity of grape marc phenolics during simulated gastrointestinal digestion (GID). Results suggest that this encapsulation method could be an effective strategy for maintaining the phenolic's antioxidant properties [101].

Curcumin is the major component of turmeric, it gives its medicinal qualities such as antioxidant abilities, anticancer, and anti-inflammatory activity [102]. It is a biologically active molecule that is typically utilized as a model for nutrition and medicine [103]. Yan et al. created nanocapsules and encapsulated curcumin using the electrostatic interaction between pectin and heat-denatured lactoferrin. With a loading capacity of 13.4%, the simulated complex demonstrated an encapsulation efficiency of almost 85.3%. The authors also concluded that the generated nanocapsules, enhanced curcumin solubility and antioxidant capacity, can be used as an appropriate delivery mechanism for sustained release [104].

To deliver the anticancer medication doxorubicin hydrochloride, researchers created a hollow nanocapsule with pectin and chitosan using a layer-by-layer technique. They reported that the nanocapsules exhibited a high loading capacity, good biocompatibility, and pH sensitivity [105].

6.5 Pectin-based nanoemulsions

Emulsions are combinations of two or more liquids that are usually immiscible in nature. The two immiscible liquids found in food are usually an oil phase and a water phase. Nanoemulsions are encapsulation systems that normally have an average diameter of between 50 and 200 nm [106]. Because of their excellent physical stability, great optical transparency, and capacity to increase the bioavailability of encapsulated drugs, nanoemulsions are frequently utilized as delivery vehicles for bioactive components [107]. Based on the relative spatial arrangement of the water and oil phases, nanoemulsions can be categorized as oil-in-water (O/W), water-in-oil (W/O), oil-in-water-in-oil (O/W/O), or water-in-oil-in-water (W/O/W) forms [108]. W/O nanoemulsions are utilized to encapsulate hydrophilic components like vitamin C, polyphenols, whereas O/W nanoemulsions are typically used to deliver hydrophobic bioactive components like carotenoids, vitamins, omega-3 fatty acids, essential oils, and curcuminoids. O/W nanoemulsions can be stabilized by polysaccharides. The polysaccharides attach themselves to the surfaces of the oil droplets and create a barrier that keeps the droplets from aggregating together. Moreover, polysaccharides can be combined with other molecules to generate conjugates that are surface-active and useful for stabilizing O/W nanoemulsions [109, 110].

Due to properties such as natural colorants and antioxidants, saffron is a chemical that is frequently employed in the food and pharmaceutical industries. It comprises bioactive constituents such as safranal, picrocrocin, and crocin. In a study, researchers created a nanoencapsulation system for saffron extract using spray-drying and double-layered emulsion utilizing whey protein concentrate and pectin, respectively. With excellent encapsulation effectiveness, these scientists were able to generate particles with a size of less than 100 nm that were devoid of pores and cracks [111].

The efficiency of low-methoxy pectin (LMP) and whey protein isolate (WPI) based complexes on the regulated release of vitamin E and vitamin B12 was investigated in a study using the double emulsion (W/O/W) method [112]. They found that the WPI–LMP complex system exhibit a 1.4-fold increase in the encapsulation efficiency of vitamin E and vitamin B12, when compared to the encapsulation efficiency of the WPI-only system. The rate of controlled release of these vitamins is also significantly improved by these complexes. The LMP shows more synergistic effects due to its diverse protein- and oil-binding capabilities. They concluded that pectin could therefore be used for nanoencapsulation, which can improve the bioaccessibility and allow for the regulated, targeted release of bioactive compounds under simulated gastric environments.

In another study, pectin-based nanoemulsions encapsulating the poorly water-soluble antifungal medication itraconazole (ITZ) were developed for use in pharmaceutical applications. They discovered that high methoxy pectin can act as a good emulsifier during the emulsion process due to its higher degree of esterification, which has a quantity of hydrophobic molecules. The discovery revealed that the molecular interaction between ITZ and pectin plays a key role in obtaining nanosized emulsion. They further suggested the possibility of using the produced nanoemulsions to develop self-emulsifying drug delivery systems [113].

6.6 Pectin-based nanoliposomes

The interactions that take place between hydrophobic polar lipid molecules, like phospholipids and hydrophilic water molecules, are typically used to generate liposomes. However, the instability of liposomes might cause changes in the distribution of particle sizes, leakage of encapsulated substances, and rapid oxidation, which limits their practical use. The generation of polymeric bio-adhesive membranes around the liposomes may help to improve this limitation [114]. Coating liposomes with a pectin biopolymer layer can improve their stability [115]. The non-toxic, biocompatible, and biodegradable nature of pectin makes it an effective choice as a coating layer for a liposome-based delivery method. Furthermore, because of its high mucoadhesive qualities and charge density, pectin can help enhance the targetability and stability of delivery systems like liposomes [116]. Zhou et al. reported that pectin-coated liposomes demonstrated the capacity to stabilize liposome-based drug delivery systems [114]. Another research has shown that pectin at a suitable concentration but with a lower degree of methylation (DM) can result in pectin-coated liposomes that are more stable and smaller in size [116].

Zhou et al. effectively created pectin-coated vitamin C encapsulated nanoliposomes in an effort to improve its stability and epidermal penetration. They discovered that low-methoxy pectin (LMP), might function as an efficient transdermal drug delivery system [114]. Pectin-based nanoliposomes were successfully generated by Haghighi et al. to encapsulate the biologically active polyphenol Phloridzin. They also observed that the pectin-coated nanoliposomes exhibit greater encapsulation effectiveness and stability when compared to non-coated nanoliposomes. Additionally, they concluded that the created pectin nanoliposomes loaded with Phloridzin could be a potential ingredient for use in both food and pharmaceutical goods [116].

Citrus fruits are a common source of Neohesperidin, a flavanone glycoside with a variety of biological actions, including anti-diabetic, anticancer, anti-inflammatory, anti-allergic, gastric protection, and neuroprotection. In order to enable sustained delivery of Neohesperidin in a simulated gastrointestinal environment, scientists created nanoliposomes by using chitosan and pectin molecules. The generated nano-carrier had a diameter of 87–225 nm. The researchers discovered that these nanoliposomes were allowed to retain 72.78% of the encapsulated substance in gastrointestinal settings and that they significantly controlled the release of Neohesperidin. They also concluded that these nanoliposomes could improve the cellular absorption of the colonic epithelial cells [117].

7. Conclusions

A new and promising avenue for the development of effective and practically useful encapsulating systems is the introduction of non-toxic, inexpensive, ecologically benign, and biocompatible biopolymers like pectin. Pectin, a ubiquitous natural biomaterial, is now an incredible matrix polymer for encapsulation. Pectin is very efficient in enhancing the bioavailability and stability of encapsulated bioactive substances. Additionally, as a molecule, it can be modified to meet certain objectives, such as controlled and targeted release of cargo and offer health advantages. Because of its potential for enhancing food quality and creating novel nutritional supplements, the technological application of these bioactive food ingredients impacts industrial development and inspires future multidisciplinary studies. To facilitate oral ingestion and more improved applications in food and pharmaceuticals, they must be used with safety requirements. Although pectin has been employed in food applications for ages, its usage in micro and nanoencapsulation is still relatively new. Future studies are expected to lead us to an advantageous situation where we can fully utilize the encapsulating potential of pectin by customizing its structures in a more evolved way on an industrial context.

Acknowledgements

We acknowledge the KSCSTE Research Fellowship, Kerala State Council for Science, Technology and Environment, Kerala, India.

Author details

O.K. Sasina Sai, Usha K. Aravind* and Charuvila T. Aravindakumar
Cochin University of Science and Technology, Kochi, India

*Address all correspondence to: uka@cusat.ac.in

References

[1] Yang M, Liang Z, Wang L, Qi M, Luo Z, Li L. Microencapsulation delivery system in food industry—Challenge and the way forward. Advances in Polymer Technology. 2020;**2020**:1-14. DOI: 10.1155/2020/7531810

[2] Quiros-Sauceda AE, Ayala-Zavala JF, Olivas GI, Gonzalez-Aguilar GA. Edible coatings as encapsulating matrices for bioactive compounds: A review. Journal of Food Science and Technology. 2014;**51**(9):1674-1685. DOI: 10.1007/s13197-013-1246-x

[3] Đorđević V, Balanč B, Belščak-Cvitanović A, Lević S, Trifković K, Kalušević A, et al. Trends in encapsulation technologies for delivery of food bioactive compounds. Food Engineering Reviews. 2014;7(4):452-490. DOI: 10.1007/s12393-014-9106-7

[4] Szliszka E, Czuba ZP, Domino M, Mazur B, Zydowicz G, Krol W. Ethanolic extract of propolis (EEP) enhances the apoptosis-inducing potential of TRAIL in cancer cells. Molecules. 2009;**14**(2):738-754. DOI: 10.3390/molecules

[5] Morales-Medina R, Drusch S, Acevedo F, Castro-Alvarez A, Benie A, Poncelet D, et al. Structure, controlled release mechanisms and health benefits of pectins as an encapsulation material for bioactive food components. Food & Function. 2022;**13**(21):10870-10881. DOI: 10.1039/d2fo00350c

[6] Chandrayan P. Biological function(s) and application (s) of pectin and pectin degrading enzymes. Biosciences, Biotechnology Research Asia. 2018;**15**(1):87-100. DOI: 10.13005/bbra/2611

[7] Voragen AGJ, Coenen G-J, Verhoef RP, Schols HA. Pectin, a versatile polysaccharide present in plant cell walls. Structural Chemistry. 2009;**20**(2):263-275. DOI: 10.1007/s11224-009-9442-z

[8] Thakur BR, Singh RK, Handa AK. Chemistry and uses of pectin–A review. Critical Reviews in Food Science and Nutrition. 1997;**37**(1):47-73. DOI: 10.1080/10408399709527767

[9] Dranca F, Oroian M. Extraction, purification and characterization of pectin from alternative sources with potential technological applications. Food Research International. 2018;**113**:327-350. DOI: 10.1016/j.foodres.2018.06.065

[10] Mollea C, Chiampo F, Conti R. Extraction and characterization of pectins from cocoa husks: A preliminary study. Food Chemistry. 2007;**107**(3):1353-1356. DOI: 10.1016/j.foodchem.2007.09.006

[11] Xu Y, Zhang L, Bailina Y, Ge Z, Ding T, Ye X, et al. Effects of ultrasound and/or heating on the extraction of pectin from grapefruit peel. Journal of Food Engineering. 2014;**126**:72-81. DOI: 10.1016/j.jfoodeng.2013.11.004

[12] Moorthy IG, Maran JP, Surya SM, Naganyashree S, Shivamathi CS. Response surface optimization of ultrasound assisted extraction of pectin from pomegranate peel. International Journal of Biological Macromolecules. 2015;**72**:1323-1328. DOI: 10.1016/j.ijbiomac.2014.10.037

[13] Kulkarni SG, Vijayanand P. Effect of extraction conditions on the quality characteristics of pectin from passion fruit peel (Passiflora edulis f. flavicarpa L.). LWT - Food Science and Technology. 2010;**43**(7):1026-1031. DOI: 10.1016/j.lwt.2009.11.006

[14] Berardini N, Knödler M, Schieber A, Carle R. Utilization of mango peels as a source of pectin and polyphenolics. Innovative Food Science & Emerging Technologies. 2005;**6**(4):442-452. DOI: 10.1016/j.ifset.2005.06.004

[15] Gopi D, Kanimozhi K, Bhuvaneshwari N, Indira J, Kavitha L. Novel banana peel pectin mediated green route for the synthesis of hydroxyapatite nanoparticles and their spectral characterization. Spectrochimica Acta. Part A, Molecular and Biomolecular Spectroscopy. 2014;**118**:589-597. DOI: 10.1016/j.saa.2013.09.034

[16] Yuliarti O, Goh KK, Matia-Merino L, Mawson J, Brennan C. Extraction and characterisation of pomace pectin from gold kiwifruit (Actinidia chinensis). Food Chemistry. 2015;**187**:290-296. DOI: 10.1016/j.foodchem.2015.03.148

[17] Chaharbaghi E, Khodaiyan F, Hosseini SS. Optimization of pectin extraction from pistachio green hull as a new source. Carbohydrate Polymers. 2017;**173**:107-113. DOI: 10.1016/j.carbpol.2017.05.047

[18] Grassino AN, Brncic M, Vikic-Topic D, Roca S, Dent M, Brncic SR. Ultrasound assisted extraction and characterization of pectin from tomato waste. Food Chemistry. 2016;**198**:93-100. DOI: 10.1016/j.foodchem.2015.11.095

[19] Abou-Elseoud WS, Hassan EA, Hassan ML. Extraction of pectin from sugar beet pulp by enzymatic and ultrasound-assisted treatments. Carbohydrate Polymer Technologies and Applications. 2021;**2**. DOI: 10.1016/j.carpta.2021.100042

[20] Miyamoto A, Chang KC. Extraction and physicochemical characterization of pectin from sunflower head residues. Journal of Food Science. 1992;**57**(6):1439-1443. DOI: 10.1111/j.1365-2621.1992.tb06878.x

[21] Košťálová Z, Aguedo M, Hromádková Z. Microwave-assisted extraction of pectin from unutilized pumpkin biomass. Chemical Engineering and Processing: Process Intensification. 2016;**102**:9-15. DOI: 10.1016/j.cep.2015.12.009

[22] Mendez DA, Fabra MJ, Gomez-Mascaraque L, Lopez-Rubio A, Martinez-Abad A. Modelling the extraction of pectin towards the valorisation of watermelon rind waste. Foods. 2021;**10**(4):738. DOI: 10.3390/foods10040738

[23] Yang JS, Mu TH, Ma MM. Extraction, structure, and emulsifying properties of pectin from potato pulp. Food Chemistry. 2018;**244**:197-205. DOI: 10.1016/j.foodchem.2017.10.059

[24] Thiraviam V, Mahejibin K. Role of Pectin in Food Processing and Food Packaging. In: Martin M, editor. Pectins. London, UK, Rijeka. p. Ch. 4: IntechOpen; 2019. DOI: 10.5772/intechopen.83677

[25] Hosseini SS, Khodaiyan F, Yarmand MS. Aqueous extraction of pectin from sour orange peel and its preliminary physicochemical properties. International Journal of Biological Macromolecules. 2016;**82**:920-926. DOI: 10.1016/j.ijbiomac.2015.11.007

[26] Zhang S, Waterhouse GIN, Xu F, He Z, Du Y, Lian Y, et al. Recent advances in utilization of pectins in biomedical applications: A review focusing on molecular structure-directing health-promoting properties. Critical Reviews in Food Science and Nutrition. 2023;**63**(19):3386-3419. DOI: 10.1080/10408398.2021.1988897

[27] Dongowski G, Lorenz A, Proll J. The degree of methylation influences the degradation of pectin in the intestinal tract of rats and in vitro. The Journal of Nutrition. 2002;**132**(7):1935-1944. DOI: 10.1093/jn/132.7.1935

[28] Chaouch MA, Hafsa J, Rihouey C, Le Cerf D, Majdoub H. Depolymerization of polysaccharides from opuntia ficus indica: Antioxidant and antiglycated activities. International Journal of Biological Macromolecules. 2015;**79**:779-786. DOI: 10.1016/j.ijbiomac.2015.06.003

[29] Wikiera A, Irla M, Mika M. Health-promoting properties of pectin. Postępy Higieny i Medycyny Doświadczalnej (Online). 2014;**68**:590-596. DOI: 10.5604/17322693.1102342

[30] Flourie B, Vidon N, Florent CH, Bernier JJ. Effect of pectin on jejunal glucose absorption and unstirred layer thickness in normal man. Gut. 1984;**25**(9):936-941. DOI: 10.1136/gut.25.9.936

[31] Dunaif G, Schneeman BO. The effect of dietary fiber on human pancreatic enzyme activity in vitro. The American Journal of Clinical Nutrition. 1981;**34**(6):1034-1035. DOI: 10.1093/ajcn/34.6.1034

[32] Di Lorenzo C, Williams CM, Hajnal F, Valenzuela JE. Pectin delays gastric emptying and increases satiety in obese subjects. Gastroenterology. 1988;**95**(5):1211-1215. DOI: 10.1016/0016-5085(88)90352-6

[33] Holt S, Carter D, Tothill P, Heading R, Prescott L. Effect of gel fibre on gastric emptying and absorption of glucose and paracetamol. The Lancet. 1979;**313**(8117):636-639. DOI: 10.1016/S0140-6736(79)91079-1

[34] Savastano DM, Hodge RJ, Nunez DJ, Walker A, Kapikian R. Effect of two dietary fibers on satiety and glycemic parameters: A randomized, double-blind, placebo-controlled, exploratory study. Nutrition Journal. 2014;**13**:45. DOI: 10.1186/1475-2891-13-45

[35] Daou C, Zhang H. Functional and physiological properties of total, soluble, and insoluble dietary fibres derived from defatted rice bran. Journal of Food Science and Technology. 2014;**51**(12):3878-3885. DOI: 10.1007/s13197-013-0925-y

[36] Munakata A, Iwane S, Todate M, Nakaji S, Sugawara K. Effects of dietary fiber on gastrointestinal transit time, fecal properties and fat absorption in rats. The Tohoku Journal of Experimental Medicine. 1995;**176**(4):227-238. DOI: 10.1620/tjem.176.227

[37] de Vries J, Miller PE, Verbeke K. Effects of cereal fiber on bowel function: A systematic review of intervention trials. World Journal of Gastroenterology. 2015;**21**(29):8952-8963. DOI: 10.3748/wjg.v21.i29.8952

[38] Kaczmarczyk MM, Miller MJ, Freund GG. The health benefits of dietary fiber: Beyond the usual suspects of type 2 diabetes mellitus, cardiovascular disease and colon cancer. Metabolism. 2012;**61**(8):1058-1066. DOI: 10.1016/j.metabol.2012.01.017

[39] Martau GA, Mihai M, Vodnar DC. The use of chitosan, alginate, and pectin in the biomedical and food sector-biocompatibility, bioadhesiveness, and biodegradability. Polymers (Basel). 2019;**11**(11):1837. DOI: 10.3390/polym11111837

[40] Sriamornsak P, Wattanakorn N, Takeuchi H. Study on the mucoadhesion mechanism of pectin by atomic force microscopy and mucin-particle method. Carbohydrate Polymers.

2010;**79**(1):54-59. DOI: 10.1016/j.carbpol.2009.07.018

[41] Gan J, Huang Z, Yu Q, Peng G, Chen Y, Xie J, et al. Microwave assisted extraction with three modifications on structural and functional properties of soluble dietary fibers from grapefruit peel. Food Hydrocolloids. 2020;**101**:105549. DOI: 10.1016/j.foodhyd.2019.105549

[42] Brouns F, Theuwissen E, Adam A, Bell M, Berger A, Mensink RP. Cholesterol-lowering properties of different pectin types in mildly hyper-cholesterolemic men and women. European Journal of Clinical Nutrition. 2012;**66**(5):591-599. DOI: 10.1038/ejcn.2011.208

[43] Theuwissen E, Mensink RP. Water-soluble dietary fibers and cardiovascular disease. Physiology & Behavior. 2008;**94**(2):285-292. DOI: 10.1016/j.physbeh.2008.01.001

[44] Miettinen TA, Tarpila S. Effect of pectin on serum cholesterol, fecal bile acids and biliary lipids in normolipidemic and hyperlipidemic individuals. Clinica Chimica Acta. 1977;**79**(2):471-477. DOI: 10.1016/0009-8981(77)90444-2

[45] Freitas CMP, Coimbra JSR, Souza VGL, Sousa RCS. Structure and applications of pectin in food, biomedical, and pharmaceutical industry: A review. Coatings. 2021;**11**(8):922. DOI: 10.3390/coatings11080922

[46] Silva DC, Freitas AL, Pessoa CD, Paula RC, Mesquita JX, Leal LK, et al. Pectin from Passiflora edulis shows anti-inflammatory action as well as hypoglycemic and hypotriglyceridemic properties in diabetic rats. Journal of Medicinal Food. 2011;**14**(10):1118-1126. DOI: 10.1089/jmf.2010.0220

[47] Umar S, Morris AP, Kourouma F, Sellin JH. Dietary pectin and calcium inhibit colonic proliferation in vivo by differing mechanisms. Cell Proliferation. 2003;**36**(6):361-375. DOI: 10.1046/j.1365-2184.2003.00291.x

[48] Liu HY, Huang ZL, Yang GH, Lu WQ, Yu NR. Inhibitory effect of modified citrus pectin on liver metastases in a mouse colon cancer model. World Journal of Gastroenterology. 2008;**14**(48):7386-7391. DOI: 10.3748/wjg.14.7386

[49] Dresler H, Keizman D, Frenkel M, Peer A, Rosenbaum E, Margel D, et al. Long term effect of PectaSol-C modified citrus pectin treatment in non-metastatic biochemically relapsed prostate cancer patients: Results of a prospective phase II study. European Urology Supplements. 2019;**18**(11):e3467. DOI: 10.1016/s1569-9056(19)34628-7

[50] Fang T, Liu DD, Ning HM, Dan L, Sun JY, Huang XJ, et al. Modified citrus pectin inhibited bladder tumor growth through downregulation of galectin-3. Acta Pharmacologica Sinica. 2018;**39**(12):1885-1893. DOI: 10.1038/s41401-018-0004-z

[51] Taper HS, Roberfroid M. Influence of inulin and oligofructose on breast cancer and tumor Growth1. The Journal of Nutrition. 1999;**129**(7):1488S-1491S. DOI: 10.1093/jn/129.7.1488S

[52] Leclere L, Fransolet M, Cote F, Cambier P, Arnould T, Van Cutsem P, et al. Heat-modified citrus pectin induces apoptosis-like cell death and autophagy in HepG2 and A549 cancer cells. PLoS One. 2015;**10**(3):e0115831. DOI: 10.1371/journal.pone.0115831

[53] Jia W, Rajani C, Xu H, Zheng X. Gut microbiota alterations are distinct for primary colorectal cancer and hepatocellular carcinoma. Protein &

Cell. 2021;**12**(5):374-393. DOI: 10.1007/s13238-020-00748-0

[54] Chung WSF, Meijerink M, Zeuner B, Holck J, Louis P, Meyer AS, et al. Prebiotic potential of pectin and pectic oligosaccharides to promote anti-inflammatory commensal bacteria in the human colon. FEMS Microbiology Ecology. 2017;**93**(11):fix127. DOI: 10.1093/femsec/fix127

[55] Sanders ME, Merenstein DJ, Reid G, Gibson GR, Rastall RA. Probiotics and prebiotics in intestinal health and disease: From biology to the clinic. Nature Reviews. Gastroenterology & Hepatology. 2019;**16**(10):605-616. DOI: 10.1038/s41575-019-0173-3

[56] Singh V, Yeoh BS, Walker RE, Xiao X, Saha P, Golonka RM, et al. Microbiota fermentation-NLRP3 axis shapes the impact of dietary fibres on intestinal inflammation. Gut. 2019;**68**(10):1801-1812. DOI: 10.1136/gutjnl-2018-316250

[57] Ishisono K, Mano T, Yabe T, Kitaguchi K. Dietary fiber pectin ameliorates experimental colitis in a neutral sugar side chain-dependent manner. Frontiers in Immunology. 2019;**10**:2979. DOI: 10.3389/fimmu.2019.02979

[58] Gottesmann M, Paraskevopoulou V, Mohammed A, Falcone FH, Hensel A. BabA and LPS inhibitors against helicobacter pylori: Pectins and pectin-like rhamnogalacturonans as adhesion blockers. Applied Microbiology and Biotechnology. 2020;**104**(1):351-363. DOI: 10.1007/s00253-019-10234-1

[59] Rehman A, Ahmad T, Aadil RM, Spotti MJ, Bakry AM, Khan IM, et al. Pectin polymers as wall materials for the nano-encapsulation of bioactive compounds. Trends in Food Science & Technology. 2019;**90**:35-46. DOI: 10.1016/j.tifs.2019.05.015

[60] Desai KGH, Jin Park H. Recent developments in microencapsulation of food ingredients. Drying Technology. 2005;**23**(7):1361-1394. DOI: 10.1081/drt-200063478

[61] Paliya BS, Sharma VK, Sharma M, Diwan D, Nguyen QD, Aminabhavi TM, et al.Protein-polysaccharidenanoconjugates: Potential tools for delivery of plant-derived nutraceuticals. Food Chemistry. 2023;**428**:136709. DOI: 10.1016/j.foodchem.2023.136709

[62] Zhang Z, Hao G, Liu C, Fu J, Hu D, Rong J, et al. Recent progress in the preparation, chemical interactions and applications of biocompatible polysaccharide-protein nanogel carriers. Food Research International. 2021;**147**:110564. DOI: 10.1016/j.foodres.2021.110564

[63] Liu C, Zhang Z, Kong Q, Zhang R, Yang X. Enhancing the antitumor activity of tea polyphenols encapsulated in biodegradable nanogels by macromolecular self-assembly. RSC Advances. 2019;**9**(18):10004-10016. DOI: 10.1039/c8ra07783e

[64] Neamtu I, Rusu AG, Diaconu A, Nita LE, Chiriac AP. Basic concepts and recent advances in nanogels as carriers for medical applications. Drug Delivery. 2017;**24**(1):539-557. DOI: 10.1080/10717544.2016.1276232

[65] Kabanov AV, Vinogradov SV. Nanogels as pharmaceutical carriers: Finite networks of infinite capabilities. Angewandte Chemie (International Ed. in English). 2009;**48**(30):5418-5429. DOI: 10.1002/anie.200900441

[66] Shishir MRI, Xie L, Sun C, Zheng X, Chen W. Advances in micro

and nano-encapsulation of bioactive compounds using biopolymer and lipid-based transporters. Trends in Food Science & Technology. 2018;**78**:34-60. DOI: 10.1016/j.tifs.2018.05.018

[67] Zhang Z, Qiu C, Li X, McClements DJ, Jiao A, Wang J, et al. Advances in research on interactions between polyphenols and biology-based nano-delivery systems and their applications in improving the bioavailability of polyphenols. Trends in Food Science & Technology. 2021;**116**:492-500. DOI: 10.1016/j.tifs.2021.08.009

[68] Sutar PB, Mishra RK, Pal K, Banthia AK. Development of pH sensitive polyacrylamide grafted pectin hydrogel for controlled drug delivery system. Journal of Materials Science. Materials in Medicine. 2008;**19**(6):2247-2253. DOI: 10.1007/s10856-007-3162-y

[69] Mishra RK, Datt M, Pal K, Banthia AK. Preparation and characterization of amidated pectin based hydrogels for drug delivery system. Journal of Materials Science. Materials in Medicine. 2008;**19**(6):2275-2280. DOI: 10.1007/s10856-007-3310-4

[70] Peng H, Chen S, Luo M, Ning F, Zhu X, Xiong H. Preparation and self-assembly mechanism of bovine serum albumin-citrus Peel pectin conjugated hydrogel: A potential delivery system for vitamin C. Journal of Agricultural and Food Chemistry. 2016;**64**(39):7377-7384. DOI: 10.1021/acs.jafc.6b02966

[71] Zhou M, Hu Q, Wang T, Xue J, Luo Y. Effects of different polysaccharides on the formation of egg yolk LDL complex nanogels for nutrient delivery. Carbohydrate Polymers. 2016;**153**:336-344. DOI: 10.1016/j.carbpol.2016.07.105

[72] Zhou M, Wang T, Hu Q, Luo Y. Low density lipoprotein/pectin complex nanogels as potential oral delivery vehicles for curcumin. Food Hydrocolloids. 2016;**57**:20-29. DOI: 10.1016/j.foodhyd.2016.01.010

[73] Jung J, Arnold RD, Wicker L. Pectin and charge modified pectin hydrogel beads as a colon-targeted drug delivery carrier. Colloids and Surfaces. B, Biointerfaces. 2013;**104**:116-121. DOI: 10.1016/j.colsurfb.2012.11.042

[74] Takei T, Sato M, Ijima H, Kawakami K. In situ gellable oxidized citrus pectin for localized delivery of anticancer drugs and prevention of homotypic cancer cell aggregation. Biomacromolecules. 2010;**11**(12):3525-3530. DOI: 10.1021/bm1010068

[75] Yu CY, Yin BC, Zhang W, Cheng SX, Zhang XZ, Zhuo RX. Composite microparticle drug delivery systems based on chitosan, alginate and pectin with improved pH-sensitive drug release property. Colloids and Surfaces. B, Biointerfaces. 2009;**68**(2):245-249. DOI: 10.1016/j.colsurfb.2008.10.013

[76] Jones OG, Lesmes U, Dubin P, McClements DJ. Effect of polysaccharide charge on formation and properties of biopolymer nanoparticles created by heat treatment of β-lactoglobulin–pectin complexes. Food Hydrocolloids. 2010;**24**(4):374-383. DOI: 10.1016/j.foodhyd.2009.11.003

[77] Kumar PV, Jain NK. Suppression of agglomeration of ciprofloxacin-loaded human serum albumin nanoparticles. AAPS PharmSciTech. 2007;**8**(1):17. DOI: 10.1208/pt0801017

[78] Verma AK, Chanchal A, Kumar A. Potential of negatively charged pectin nanoparticles encapsulating paclitaxel: Preparation & Characterization. In: International Conference on Nanoscience, Technology and Societal Implications,

8-10 December 2011, Bhubaneswar, India. New York: IEEE; 2011. pp. 1-8

[79] Huang X, Dai Y, Cai J, Zhong N, Xiao H, McClements DJ, et al. Resveratrol encapsulation in core-shell biopolymer nanoparticles: Impact on antioxidant and anticancer activities. Food Hydrocolloids. 2017;**64**:157-165. DOI: 10.1016/j.foodhyd.2016.10.029

[80] Prezotti FG, Boni FI, Ferreira NN, Silva DS, Almeida A, Vasconcelos T, et al. Oral nanoparticles based on gellan gum/pectin for colon-targeted delivery of resveratrol. Drug Development and Industrial Pharmacy. 2020;**46**(2):236-245. DOI: 10.1080/03639045.2020.1716374

[81] Coelho B, Mazzarino L, Pitz HS, Feltrin C, Voytena APL, Coelho DS, et al. Development of nanoparticles coated with cassava bagasse pectin (Manihot esculenta Crantz) containing beta-carotene for mucoadhesive applications. Anais da Academia Brasileira de Ciências. 2020;**92**(4):e20200134. DOI: 10.1590/0001-3765202020190134

[82] Gan C, Liu Q, Zhang Y, Shi T, He W-S, Jia C. A novel phytosterols delivery system based on sodium caseinate-pectin soluble complexes: Improving stability and bioaccessibility. Food Hydrocolloids. 2022;**124**. DOI: 10.1016/j.foodhyd.2021.107295

[83] Hu K, Huang X, Gao Y, Huang X, Xiao H, McClements DJ. Core-shell biopolymer nanoparticle delivery systems: Synthesis and characterization of curcumin fortified zein-pectin nanoparticles. Food Chemistry. 2015;**182**:275-281. DOI: 10.1016/j.foodchem.2015.03.009

[84] Cho H, Jung H, Lee H, Kwak HK, Hwang KT. Formation of electrostatic complexes using sodium caseinate with high-methoxyl pectin and carboxymethyl cellulose and their application in stabilisation of curcumin. International Journal of Food Science & Technology. 2016;**51**(7):1655-1665. DOI: 10.1111/ijfs.13137

[85] Veneranda M, Hu Q, Wang T, Luo Y, Castro K, Madariaga JM. Formation and characterization of zein-caseinate-pectin complex nanoparticles for encapsulation of eugenol. LWT. 2018;**89**:596-603. DOI: 10.1016/j.lwt.2017.11.040

[86] Ramalingam M, Ramakrishna S. Introduction to nanofiber composites. In: Ramalingam M, Ramakrishna S, editors. Nanofiber Composites for Biomedical Applications. United Kingdom: Woodhead Publishing; 2017. pp. 3-29. DOI: 10.1016/B978-0-08-100173-8.00001-6

[87] Jain R, Shetty S, Yadav KS. Unfolding the electrospinning potential of biopolymers for preparation of nanofibers. Journal of Drug Delivery Science and Technology. 2020;**57**:101604. DOI: 10.1016/j.jddst.2020.101604

[88] Mendes AC, Gorzelanny C, Halter N, Schneider SW, Chronakis IS. Hybrid electrospun chitosan-phospholipids nanofibers for transdermal drug delivery. International Journal of Pharmaceutics. 2016;**510**(1):48-56. DOI: 10.1016/j.ijpharm.2016.06.016

[89] Stijnman AC, Bodnar I, Hans Tromp R. Electrospinning of food-grade polysaccharides. Food Hydrocolloids. 2011;**25**(5):1393-1398. DOI: 10.1016/j.foodhyd.2011.01.005

[90] Tomasula PM, Sousa AMM, Liou SC, Li R, Bonnaillie LM, Liu LS. Short communication: Electrospinning of casein/pullulan blends for food-grade applications. Journal of Dairy Science. 2016;**99**(3):1837-1845. DOI: 10.3168/jds.2015-10374

[91] Cui S, Yao B, Gao M, Sun X, Gou D, Hu J, et al. Effects of pectin structure and crosslinking method on the properties of crosslinked pectin nanofibers. Carbohydrate Polymers. 2017;**157**:766-774. DOI: 10.1016/j.carbpol.2016.10.052

[92] Alborzi S, Lim LT, Kakuda Y. Encapsulation of folic acid and its stability in sodium alginate-pectin-poly(ethylene oxide) electrospun fibres. Journal of Microencapsulation. 2013;**30**(1):64-71. DOI: 10.3109/02652048.2012.696153

[93] Cui S, Yao B, Sun X, Hu J, Zhou Y, Liu Y. Reducing the content of carrier polymer in pectin nanofibers by electrospinning at low loading followed with selective washing. Materials Science & Engineering. C, Materials for Biological Applications. 2016;**59**:885-893. DOI: 10.1016/j.msec.2015.10.086

[94] Cui SS, Sun X, Yao B, Peng XX, Zhang XT, Zhou YF, et al. Size-tunable low molecular weight pectin-based electrospun nanofibers blended with low content of poly(ethylene oxide). Journal of Nanoscience and Nanotechnology. 2017;**17**(1):681-689. DOI: 10.1166/jnn.2017.12540

[95] Jiang B, Hu L, Gao C, Shen J. Crosslinked polysaccharide nanocapsules: Preparation and drug release properties. Acta Biomaterialia. 2006;**2**(1):9-18. DOI: 10.1016/j.actbio.2005.08.006

[96] Alkanawati MS, da Costa Marques R, Mailander V, Landfester K, Therien-Aubin H. Polysaccharide-based pH-responsive nanocapsules prepared with bio-orthogonal chemistry and their use as responsive delivery systems. Biomacromolecules. 2020;**21**(7):2764-2771. DOI: 10.1021/acs.biomac.0c00492

[97] Yan X, Chai L, Fleury E, Ganachaud F, Bernard J. 'Sweet as a nut': Production and use of nanocapsules made of glycopolymer or polysaccharide shell. Progress in Polymer Science. 2021;**120**. DOI: 10.1016/j.progpolymsci.2021.101429

[98] Huang Y, Lu Z, Liu F, Lane J, Chen J, Fu X, et al. Osteopontin enhances the probiotic viability of bifidobacteria in pectin-based microencapsulation subjected to in vitro infant gastrointestinal digestion. Food Hydrocolloids. 2024;**149**. DOI: 10.1016/j.foodhyd.2023.109634

[99] Jiang R, Prell C, Lonnerdal B. Milk osteopontin promotes brain development by up-regulating osteopontin in the brain in early life. The FASEB Journal. 2019;**33**(2):1681-1694. DOI: 10.1096/fj.201701290RR

[100] He C, Sampers I, Van de Walle D, Dewettinck K, Raes K. Encapsulation of lactobacillus in low-methoxyl pectin-based microcapsules stimulates biofilm formation: Enhanced resistances to heat shock and simulated gastrointestinal digestion. Journal of Agricultural and Food Chemistry. 2021;**69**(22):6281-6290. DOI: 10.1021/acs.jafc.1c00719

[101] Cruz-Molina AV, Goncalves C, Neto MD, Pastrana L, Jauregi P, Amado IR. Whey-pectin microcapsules improve the stability of grape marc phenolics during digestion. Journal of Food Science. 2023;**88**(12):4892-4906. DOI: 10.1111/1750-3841.16806

[102] Malaekeh-Nikouei B, Salarbashi D. Nanoencapsulation and delivery of curcumin using some carbohydrate based systems: A review. Nanomedicine Journal. 2018;**5**(2):57-61. DOI: 10.22038/nmj.2018.005.001

[103] Arpagaus C, Collenberg A, Rutti D, Assadpour E, Jafari SM. Nano spray drying for encapsulation of

pharmaceuticals. International Journal of Pharmaceutics. 2018;**546**(1-2):194-214. DOI: 10.1016/j.ijpharm.2018.05.037

[104] Yan JK, Qiu WY, Wang YY, Wu JY. Biocompatible polyelectrolyte complex nanoparticles from Lactoferrin and pectin as potential vehicles for antioxidative curcumin. Journal of Agricultural and Food Chemistry. 2017;**65**(28):5720-5730. DOI: 10.1021/acs.jafc.7b01848

[105] Ji F, Li J, Qin Z, Yang B, Zhang E, Dong D, et al. Engineering pectin-based hollow nanocapsules for delivery of anticancer drug. Carbohydrate Polymers. 2017;**177**:86-96. DOI: 10.1016/j.carbpol.2017.08.107

[106] Araiza-Calahorra A, Akhtar M, Sarkar A. Recent advances in emulsion-based delivery approaches for curcumin: From encapsulation to bioaccessibility. Trends in Food Science & Technology. 2018;**71**:155-169. DOI: 10.1016/j.tifs.2017.11.009

[107] Salvia-Trujillo L, McClements DJ. Influence of Nanoemulsion addition on the stability of conventional emulsions. Food Biophysics. 2015;**11**(1):1-9. DOI: 10.1007/s11483-015-9401-8

[108] Bhushani A, Anandha-ramakrishnan C. Food-grade nanoemulsions for protection and delivery of nutrients. In: Ranjan S, Dasgupta N, Lichtfouse E, editors. Nanoscience in Food and Agriculture. Cham: Springer International Publishing; 2017. pp. 99-139. DOI: 10.1007/978-3-319-53112-0_3

[109] Xu D, Yuan F, Gao Y, Panya A, McClements DJ, Decker EA. Influence of whey protein-beet pectin conjugate on the properties and digestibility of beta-carotene emulsion during in vitro digestion. Food Chemistry. 2014;**156**:374-379. DOI: 10.1016/j.foodchem.2014.02.019

[110] Sun J, Liu T, Mu Y, Jing H, Obadi M, Xu B. Enhancing the stabilization of β-carotene emulsion using ovalbumin-dextran conjugates as emulsifier. Colloids and Surfaces A: Physicochemical and Engineering Aspects. 2021;**626**. DOI: 10.1016/j.colsurfa.2021.126806

[111] Esfanjani AF, Jafari SM, Assadpoor E, Mohammadi A. Nano-encapsulation of saffron extract through double-layered multiple emulsions of pectin and whey protein concentrate. Journal of Food Engineering. 2015;**165**:149-155. DOI: 10.1016/j.jfoodeng.2015.06.022

[112] Li B, Jiang Y, Liu F, Chai Z, Li Y, Li Y, et al. Synergistic effects of whey protein–polysaccharide complexes on the controlled release of lipid-soluble and water-soluble vitamins in W1/O/W2 double emulsion systems. International Journal of Food Science & Technology. 2011;**47**(2):248-254. DOI: 10.1111/j.1365-2621.2011.02832.x

[113] Burapapadh K, Kumpugdee-Vollrath M, Chantasart D, Sriamornsak P. Fabrication of pectin-based nanoemulsions loaded with itraconazole for pharmaceutical application. Carbohydrate Polymers. 2010;**82**(2):384-393. DOI: 10.1016/j.carbpol.2010.04.071

[114] Zhou W, Liu W, Zou L, Liu W, Liu C, Liang R, et al. Storage stability and skin permeation of vitamin C liposomes improved by pectin coating. Colloids and Surfaces. B, Biointerfaces. 2014;**117**:330-337. DOI: 10.1016/j.colsurfb.2014.02.036

[115] Chun J-Y, Choi M-J, Min S-G, Weiss J. Formation and stability of multiple-layered liposomes by layer-by-layer electrostatic deposition of biopolymers. Food Hydrocolloids. 2013;**30**(1):249-257. DOI: 10.1016/j.foodhyd.2012.05.024

[116] Haghighi M, Yarmand MS, Emam-Djomeh Z, McClements DJ, Saboury AA, Rafiee-Tehrani M. Design and fabrication of pectin-coated nanoliposomal delivery systems for a bioactive polyphenolic: Phloridzin. International Journal of Biological Macromolecules. 2018;**112**:626-637. DOI: 10.1016/j.ijbiomac.2018.01.108

[117] Shishir MRI, Karim N, Gowd V, Xie J, Zheng X, Chen W. Pectin-chitosan conjugated nanoliposome as a promising delivery system for neohesperidin: Characterization, release behavior, cellular uptake, and antioxidant property. Food Hydrocolloids. 2019;**95**:432-444. DOI: 10.1016/j.foodhyd.2019.04.059

Chapter 4

Biomaterials-Based Hydrogels for Therapeutic Applications

Mariana Chelu and Adina Magdalena Musuc

Abstract

Conventional therapeutic models based on the premise of a universal solution are facing a decrease in efficiency, emphasized by the large number of patients who show resistance or who do not respond positively to classic treatments. This perspective highlights the urgency for more precise approaches based on personalized treatments that are adaptable to the specific complexities and unique challenges faced by each patient. Hydrogels are biocompatible and biodegradable systems for well-controlled and targeted administration of therapeutic agents, being formed by 3D reticulated networks of water-soluble polymeric biomaterials, of natural, synthetic, or hybrid origin, with specific intrinsic and extrinsic properties. Due to the easily adjustable porous structure, hydrogels allow the encapsulation of macromolecular drugs, proteins, small molecules, cells, hormones, or growth factors in the gel matrix and their subsequent controlled release. The biomaterials used, the crosslinking methods, the design, and the functionalization strategies in obtaining hydrogels with improved properties are presented. The different possibilities of application are described transdermally, as dressing materials, oral, ocular, spray-able, or injectable, up to the intracellular level. This chapter extensively investigates the advances and unique advantages of hydrogels that enable effective, noninvasive, personalized treatments and provide greater patient comfort for a wide range of applications.

Keywords: biomaterials, hydrogels, microencapsulation, biomedical applications, bioplatforms

1. Introduction

In recent years, biomaterials-based hydrogels have emerged as promising platforms for a wide range of therapeutic applications, revolutionizing the landscape of modern medicine. These versatile materials, composed of water-swollen polymer networks, possess unique properties that make them ideal candidates for drug delivery, tissue engineering, regenerative medicine, and diagnostic applications [1–5]. Biomaterials-based hydrogels offer several distinct advantages for therapeutic applications. Their inherent biocompatibility and ability to interact with biological systems without eliciting adverse reactions make them well-suited for *in vivo* use. Additionally, hydrogels provide a versatile platform for controlled drug delivery, enabling precise modulation of drug release kinetics and spatiotemporal control over therapeutic delivery. In this era of precision medicine, biomaterials-based hydrogels hold immense promise for targeted therapy and personalized treatment approaches [6]. By incorporating bioactive molecules, growth factors, or cell-adhesive peptides into hydrogel matrices, researchers can create tailored environments that promote tissue regeneration, repair, and remodeling. Furthermore, advancements in 3D bio-printing technology enable the fabrication of complex, biomimetic tissue engineering constructs with precise control over scaffold architecture and cell distribution, offering innovative solutions for organ transplantation, wound healing, and disease modeling [7]. Despite the significant progress made in the field of biomaterials-based hydrogels, several challenges and opportunities for advancement remain. In this context, this chapter aims to provide a comprehensive overview of biomaterials-based hydrogels for therapeutic applications, exploring their design principles, fabrication techniques, biomedical applications, trends, and future prospects.

2. Introduction to biomaterials: bridging science and medicine

Biomaterials, at the intersection of science and medicine, represent a dynamic field of study and innovation with profound implications for healthcare and human well-being. These materials, engineered to interact with biological systems, hold the promise of revolutionizing medical treatments, diagnostics, and therapeutic interventions. At its core, a biomaterial is any substance that interacts with biological systems to produce a desired therapeutic or diagnostic effect. This broad definition encompasses a wide range of materials, from metals and ceramics to polymers and composites, each uniquely tailored to fulfill specific biomedical functions [6]. Biomaterials serve as the building blocks of medical devices, implants, drug delivery systems, tissue engineering scaffolds, and diagnostic tools, enabling advancements in regenerative medicine, personalized healthcare, and disease management (**Figure 1**).

Biomaterials exhibit several key characteristics that distinguish them from conventional materials:

a. Biocompatibility: Biomaterials must be biocompatible, meaning they are well-tolerated by the body without eliciting adverse immune responses or toxic reactions. Biocompatibility ensures that biomaterials can interact with biological tissues and fluids without causing harm, making them suitable for medical applications [5].

b. Bioactivity: Some biomaterials possess inherent bioactivity, meaning they can interact with biological molecules or promote specific cellular responses. Bioactive biomaterials stimulate tissue regeneration, cell adhesion, or molecular signaling, facilitating healing and repair processes within the body [6].

c. Mechanical properties: Biomaterials must possess appropriate mechanical properties to withstand physiological forces and maintain structural integrity *in vivo*. These properties vary depending on the intended application, with materials ranging from rigid metals for load-bearing implants to flexible polymers for tissue engineering scaffolds [8].

d. Degradation kinetics: Some biomaterials are designed to degrade over time, either through enzymatic degradation or hydrolytic cleavage of polymer chains.

Figure 1.
Diversity, properties, and applications of biomaterials-based hydrogels.

Degradable biomaterials enable temporary support or controlled release of therapeutic agents, followed by gradual integration or resorption within the body [2].

e. Surface modifications: Biomaterial surfaces can be modified to enhance biocompatibility, bioactivity, or functionality. Surface modifications may involve coating with bioactive molecules, tethering of cell-adhesive peptides, or incorporation of drug-releasing nanoparticles, enabling tailored interactions with biological systems [3].

The applications of biomaterials span a wide range of medical fields and disciplines, including:

Implantable medical devices: Biomaterials are used to fabricate orthopedic implants, cardiovascular stents, dental prosthetics, and neural electrodes, providing structural support, mechanical stability, and biocompatibility for long-term implantation within the body [8–10].

Tissue engineering and regenerative medicine: Biomaterials serve as scaffolds for growing tissues and organs *in vitro*, enabling tissue regeneration and repair *in vivo*. Tissue-engineered constructs, composed of biomaterials seeded with cells and growth factors, offer promising approaches for treating injuries, defects, and degenerative diseases [11, 12].

Drug delivery systems: Biomaterials are utilized to design drug delivery vehicles, such as nano/microparticles, hydrogels, and implants, enabling targeted delivery, controlled release, and enhanced bioavailability of therapeutic agents for treating various diseases, including cancer, infectious diseases, and chronic conditions [13].

Diagnostic tools and biosensors: Biomaterials are employed to develop diagnostic assays, imaging contrast agents, and biosensors for detecting biomarkers, pathogens, and disease states. Functionalized biomaterials enable sensitive and specific detection of molecular targets, facilitating early diagnosis, disease monitoring, and personalized medicine [14].

Table 1 summarizes the principal characteristics of biomaterials.

2.1 Importance of biomaterials in medicine

The importance of biomaterials in medicine cannot be overstated, as they play a pivotal role in advancing healthcare in numerous ways. Here are some key reasons why biomaterials are indispensable in the field of medicine:

1. **Tissue engineering and regenerative medicine**: Biomaterials serve as scaffolds and matrices for the regeneration and repair of damaged tissues and organs. They provide structural support and cues for cell growth, differentiation, and tissue formation, enabling the development of novel therapies for conditions ranging from bone defects to organ failure [5].

2. **Medical implants and devices**: Biomaterials are essential components of medical implants and devices used for various purposes, including joint replacements, cardiovascular stents, dental implants, and prosthetic limbs. These implants restore function, alleviate pain, and improve the quality of life for millions of patients worldwide [8].

Characteristic	Description
Biocompatibility	Ability to interact with biological systems without eliciting adverse reactions
Mechanical strength	Resistance to deformation or fracture under applied loads
Degradation kinetics	Rate of breakdown or dissolution over time
Surface topography	Physical features at the material interface, influencing cell adhesion and response
Porosity	Pore size and distribution affecting fluid transport, cell infiltration, and tissue ingrowth
Bioactivity	Ability to induce specific cellular responses or tissue regeneration
Tunable properties	Flexibility to adjust mechanical, chemical, or biological properties for specific applications
Sterility	Freedom from microbial contamination or potential for sterilization
Immunogenicity	Likelihood of eliciting an immune response *in vivo*
Stability	Maintenance of structural and functional integrity over time

Table 1.
Characteristics of biomaterials.

3. **Drug delivery systems**: Biomaterials are employed in drug delivery systems to enhance the efficacy, safety, and targeted delivery of pharmaceutical agents. By encapsulating drugs within biocompatible carriers, biomaterials enable controlled release, prolonged circulation, and site-specific targeting, thereby optimizing therapeutic outcomes and minimizing side effects [2, 11].

4. **Diagnostic and therapeutic imaging**: Biomaterials are utilized in contrast agents and imaging probes for diagnostic imaging techniques such as magnetic resonance imaging (MRI), computed tomography (CT), and ultrasound. These biomaterial-based imaging agents enable enhanced visualization of anatomical structures, pathological changes, and molecular markers, facilitating early detection, accurate diagnosis, and treatment monitoring [15].

5. **Biosensors and diagnostics**: Biomaterials are integral components of biosensors and diagnostic assays for detecting biomolecules, pathogens, and physiological parameters with unique properties, such as biorecognition, signal transduction, and amplification capabilities. Biosensors enable rapid, sensitive, and specific detection of disease biomarkers [14].

6. **Surgical and wound care products**: Biomaterials are utilized in surgical sutures, tissue adhesives, and wound dressings to promote hemostasis, wound closure, and tissue healing [16, 17]. These biomaterial-based products enhance surgical outcomes, reduce complications, and accelerate the healing process, improving patient recovery and reducing healthcare costs.

7. **Biocompatibility testing and research**: Biomaterials serve as valuable tools for studying cell–material interactions, biocompatibility, and tissue responses *in vitro* and *in vivo* [5].

8. **Personalized medicine and theragnostic**: Biomaterials are driving innovations in personalized medicine and theragnostic, which combine diagnostics and therapy into integrated platforms [18, 19]. By incorporating biomaterial-based drug delivery systems with diagnostic imaging modalities and targeted therapies, theranostic approaches enable precise diagnosis, treatment monitoring, and therapeutic interventions tailored to individual patients' needs.

Table 2 provides a basic overview of the advantages and disadvantages of using biomaterials in medicine.

2.2 Overview of different types of biomaterials

Biomaterials encompass a diverse array of materials engineered to interact with biological systems [20–22]. From synthetic polymers to natural substances, biomaterials offer a rich tapestry of options for addressing various medical challenges.

1. *Synthetic polymers* offer unparalleled versatility, enabling precise control over mechanical, chemical, and biological properties. Examples include polyethylene glycol (PEG), polylactic acid (PLA), and poly(lactic-*co*-glycolic acid) (PLGA). Synthetic polymers find applications in drug delivery systems, tissue engineering

Aspects	Advantages	Disadvantages
Biocompatibility	Interact well with biological systems Minimal risk of adverse reactions	Risk of immunogenicity Potential for foreign body response
Mechanical properties	Tunable mechanical properties for various applications Mimic natural tissues and organs	May lack sufficient strength or stiffness for certain load-bearing needs
Degradation kinetics	Biodegradable options available for temporary applications Controlled degradation can facilitate tissue regeneration	Variable degradation rates may affect longevity of implants Potential for inflammatory response during degradation process
Surface topography	Tailorable surface properties for specific applications Influence cell adhesion, proliferation, and differentiation	Surface modifications may alter biocompatibility or functionality
Porosity	Porous structures facilitate nutrient and waste exchange Support cell infiltration and tissue ingrowth	May compromise structural integrity or mechanical properties
Bioactivity	Induce specific cellular responses or tissue regeneration	Potential for unintended biological effects or off-target interactions
Tunability	Flexibility to adjust properties for diverse medical needs	Requires careful optimization and characterization
Sterility	Potential for sterilization to minimize infection risk	Sterilization methods may alter material properties or structure

Table 2.
Advantages and disadvantages of using biomaterials in medicine.

scaffolds, and medical implants, where their tailored functionalities facilitate targeted therapeutic interventions and tissue regeneration [23, 24].

2. *Natural polymers* derived from biological sources such as proteins, polysaccharides, and extracellular matrices, natural polymers offer inherent biocompatibility and bioactivity. Collagen, derived from connective tissues, serves as a ubiquitous scaffold for tissue engineering and wound healing. Hyaluronic acid, a glycosaminoglycan found in the extracellular matrix, exhibits excellent lubricating and viscoelastic properties, making it ideal for ophthalmic and orthopedic applications. Natural polymers offer the advantage of mimicking the complexity and biochemical cues of native tissues, fostering cell adhesion, proliferation, and differentiation in regenerative medicine approaches [25, 26].

3. *Metals and Alloys* renowned for their strength, durability, and biocompatibility, are indispensable in orthopedic and dental implants, cardiovascular stents, and surgical instruments. Titanium and its alloys, such as Ti-6Al-4V, are prized for their excellent corrosion resistance and osseointegration properties, making them ideal for load-bearing implants. Stainless steel, with its high strength and machinability, finds applications in orthopedic fixation devices and cardiovascular stents. Metals and alloys provide structural support and mechanical stability, ensuring the long-term success of implantable medical devices [27, 28].

4. *Ceramics* characterized by their hardness, biocompatibility, and bioactivity, hold promise for various biomedical applications. Hydroxyapatite, a calcium phosphate ceramic, mimics the mineral composition of bone and serves as a bone graft substitute for promoting osteogenesis and osseointegration. Bioactive glasses, such as silicate-based compositions, exhibit osteoconductive and angiogenic properties, facilitating bone regeneration and tissue repair. Ceramics offer excellent biocompatibility and stability in physiological environments, making them valuable materials for bone substitutes, dental restorations, and tissue engineering scaffolds [29, 30].

5. *Composites* composed of two or more distinct materials combine the advantageous properties of each component to achieve synergistic effects and to address specific biomedical needs [31, 32]. For example, polymer-ceramic composites merge the flexibility of polymers with the strength and bioactivity of ceramics, offering enhanced mechanical properties and biological performance for bone tissue engineering and dental restorations. Carbon nanotube-reinforced polymers harness the exceptional mechanical strength and electrical conductivity of carbon nanotubes for applications in neural interfaces and tissue engineering scaffolds.

3. Hydrogels

Hydrogels, a fascinating class of biomaterials, hold immense promise for a wide range of biomedical applications. These unique materials, characterized by their high-water content and tunable properties, offer versatility and biocompatibility unmatched by traditional materials. At their core, hydrogels are three-dimensional networks of hydrophilic polymer chains that swell in water but maintain their structural integrity. This unique characteristic endows hydrogels with a sponge-like quality, allowing them to absorb and retain large amounts of water while retaining their shape and mechanical properties. Hydrogels can be synthesized from both natural and synthetic polymers, offering a wide range of options for tailoring their properties to suit specific applications [33, 34].

3.1 Properties of hydrogels

1. **High-water content**: This property enables hydrogels to mimic the physiological environment, making them ideal candidates for tissue engineering and drug delivery [1].

2. **Tunable mechanical properties of hydrogels**: These can be precisely engineered by adjusting factors such as polymer composition, crosslinking density, and network architecture. This tunability allows hydrogels to mimic the mechanical properties of various tissues, ranging from soft brain tissue to stiff cartilage, making them suitable for a wide range of applications in regenerative medicine and soft tissue engineering [3].

3. **Biocompatibility and bioactivity**: Hydrogels exhibit excellent biocompatibility, providing a supportive environment for cell growth, proliferation, and tissue regeneration. Furthermore, hydrogels can be engineered to incorporate bioactive

molecules such as growth factors, peptides, and cytokines, enhancing their ability to promote tissue healing and regeneration [34].

4. **Swelling and degradation kinetics**: Hydrogels swell in response to water uptake, allowing for controlled release of encapsulated drugs or bioactive molecules. Additionally, hydrogels can be designed to degrade over time, either through enzymatic degradation or hydrolytic cleavage of polymer chains. This degradation kinetics can be tailored to match the rate of tissue regeneration, enabling transient support followed by gradual integration into native tissue [33].

5. **Responsive behavior**: Some hydrogels exhibit responsive behavior to external stimuli such as pH, temperature, and light. These stimuli-responsive hydrogels undergo reversible changes in swelling, mechanical properties, or drug release in response to changes in their environment. This property enables precise control over hydrogel behavior and functionality, opening up new avenues for smart drug delivery systems and tissue engineering scaffolds [6, 26].

3.2 Classification of hydrogels

One key aspect that distinguishes hydrogels is their classification based on their origin and composition.

1. *Natural hydrogels* are derived from biopolymers found in living organisms, offering inherent biocompatibility and bioactivity [33, 34]. These hydrogels are often sourced from proteins, polysaccharides, and extracellular matrices, which provide a natural scaffold for cellular activities. Examples of natural hydrogels include:

   - **Collagen hydrogels**: Derived from collagen, the main structural protein in connective tissues, collagen hydrogels closely mimic the composition and properties of native extracellular matrices. These hydrogels offer excellent biocompatibility and promote cell adhesion, proliferation, and tissue regeneration, making them valuable materials for wound healing, tissue engineering, and drug delivery applications [35].

   - **Alginate hydrogels**: Alginate, extracted from brown seaweed, forms hydrogels through ionic crosslinking with divalent cations such as calcium ions. Alginate hydrogels exhibit high-water content and tunable mechanical properties, making them suitable for applications in cell encapsulation, drug delivery, and tissue engineering [36].

   - **Hyaluronic acid hydrogels**: Hyaluronic acid, a glycosaminoglycan found in the extracellular matrix, forms hydrogels with excellent viscoelastic properties and biocompatibility. Hyaluronic acid hydrogels are widely used in ophthalmic and orthopedic applications, as well as in skin care products and drug delivery systems [37].

Natural hydrogels offer the advantage of mimicking the complexity and biochemical cues of native tissues, fostering cell adhesion, proliferation, and differentiation in regenerative medicine approaches.

2. *Synthetic hydrogels* are engineered from synthetic polymers and designed to exhibit tailored functionalities and mechanical properties for specific biomedical applications. Examples of synthetic hydrogels include:

   - **Polyethylene glycol (PEG) hydrogels**: PEG hydrogels are synthesized from polyethylene glycol, a biocompatible and inert polymer. These hydrogels offer tunable mechanical properties, swelling behavior, and degradation kinetics, making them versatile platforms for drug delivery, tissue engineering, and biosensing applications [24, 38].

   - **Poly(N-isopropylacrylamide) (PNIPAAm) hydrogels**: PNIPAAm hydrogels exhibit thermoresponsive behavior, undergoing reversible phase transition in response to changes in temperature. Below a critical temperature (lower critical solution temperature, LCST), PNIPAAm hydrogels swell in water, while above the LCST, they collapse and disappear. This unique property enables PNIPAAm hydrogels to be utilized in smart drug delivery systems and tissue engineering scaffolds [39].

   - **Polyacrylamide hydrogels**: Polyacrylamide hydrogels offer high mechanical strength and stability, making them suitable for applications requiring load-bearing capabilities in tissue engineering, cell culture, and microfluidic devices due to their biocompatibility and ease of fabrication [40].

3. *Hybrid hydrogels* combine elements of both natural and synthetic polymers, offering synergistic properties and functionalities. These hybrid materials leverage the advantages of natural polymers, such as biocompatibility and bioactivity, while incorporating the tunability and mechanical strength of synthetic polymers. Examples of hybrid hydrogels include [41]:

   - **Gelatin-methacryloyl (GelMA) hydrogels** are synthesized by chemically modifying gelatin with methacryloyl groups, enabling photocrosslinking and tunable mechanical properties. They combine the biocompatibility of gelatin with the versatility of photopolymerization, making them valuable materials for tissue engineering, 3D bioprinting, and drug delivery applications [7].

   - **Chitosan-polyethylene glycol (CS-PEG) hydrogels**: CS-PEG hydrogels combine the biocompatibility of chitosan with the tunability of polyethylene glycol, offering a versatile platform for drug delivery and tissue engineering. They exhibit controlled release of encapsulated drugs and growth factors, promoting tissue regeneration and wound healing [17, 38].

   - **Cellulose-based hydrogels**: Cellulose-based hydrogels incorporate cellulose derivatives with synthetic polymers to create hybrid materials with enhanced mechanical properties and biocompatibility. These hydrogels find applications in wound dressings, drug delivery systems, and tissue engineering scaffolds, leveraging the biodegradability and abundance of cellulose [33].

Hybrid hydrogels harness the synergy between natural and synthetic components, offering a platform for multifunctional and tailored materials with enhanced properties and performance. Natural hydrogels leverage the biocompatibility and bioactivity of biopolymers sourced from living organisms, while synthetic hydrogels offer precise control over properties and functionalities through engineered polymers.

3.3 Factors influencing the design and properties of hydrogels

In the realm of biomaterials, hydrogels stand out as versatile and promising materials with diverse applications in medicine and biotechnology. The design and properties of hydrogels are influenced by numerous factors, ranging from their composition and synthesis methods to environmental conditions and intended applications.

3.3.1 Composition of hydrogels

The composition of hydrogels serves as the foundation for their properties and functionalities, dictating their biocompatibility, mechanical strength, and responsiveness to stimuli. Hydrogels can be composed of natural polymers, synthetic polymers, or hybrid combinations thereof, each offering unique advantages and challenges. By leveraging the advantages of natural polymers (e.g., biocompatibility) with the tunability of synthetic polymers (e.g., mechanical strength), hybrid hydrogels provide versatile platforms for biomedical applications [42, 43].

3.3.2 Crosslinking methods

The crosslinking method, which is essentially the gel structure forming reaction, employed during hydrogel synthesis plays a critical role in determining the network architecture, mechanical properties, and responsiveness of hydrogels. Various crosslinking strategies, including physical, chemical, and biological methods, offer distinct advantages and limitations for tuning hydrogel properties [44]. Chemical crosslinking methods require precise control of reaction conditions, stoichiometry, and reaction kinetics to achieve desired crosslink densities and network structures. Physical crosslinking techniques involve optimization of environmental parameters to induce gelation [45].

- **Physical Crosslinking**: Physical crosslinking methods, such as temperature-induced gelation, ionotropic gelation, and self-assembly, rely on noncovalent interactions to form the hydrogel network. These methods offer simplicity and reversibility, enabling controlled gelation and responsive behavior, as well as lower mechanical strength and stability compared to chemically crosslinked hydrogels.

- **Chemical Crosslinking**: Chemical crosslinking methods, such as photopolymerization, Michael addition, and click chemistry, involve the formation of covalent bonds between polymer chains to create a stable hydrogel network. These methods offer precise control over crosslink density and mechanical properties, facilitating the customization of hydrogel stiffness, degradation kinetics, and drug release profiles.

- **Biological Crosslinking**: Biological crosslinking methods, such as enzymatic crosslinking and cell-mediated gelation, harness biological agents to induce hydrogel formation and remodeling. These methods offer compatibility with living cells and tissues, enabling *in situ* gelation and dynamic interactions with the biological environment.

3.3.3 Environmental conditions

The properties and behavior of hydrogels are influenced by environmental conditions such as temperature, pH, and ionic strength, which can trigger changes in swelling, mechanical properties, and drug release kinetics [46].

- **pH Responsiveness**: pH-responsive hydrogels, such as poly(acrylic acid) (PAA) hydrogels, exhibit changes in swelling behavior and mechanical properties in response to variations in pH. By incorporating pH-sensitive moieties into the hydrogel network, researchers can design hydrogels that respond to acidic or alkaline environments, enabling applications in drug delivery, biosensing, and wound healing.

- **Ionic strength and solvent composition**: Hydrogel swelling, and network properties can be influenced by variations in ionic strength and solvent composition. By adjusting the concentration of ions or solvent polarity, researchers can modulate hydrogel swelling behavior, drug release kinetics, and mechanical properties, offering opportunities for tailored hydrogel design and optimization.

4. Biomaterials for drug delivery

The convergence of biomaterials science and pharmaceuticals has given rise to a revolutionary approach in healthcare: drug delivery systems. These systems utilize biomaterials as carriers to transport therapeutic agents to specific targets within the body, offering precise control over drug release kinetics, bioavailability, and therapeutic efficacy [2, 47–50].

Biomaterial-based drug delivery systems are engineered to overcome biological barriers that limit the efficacy of conventional drug formulations. Nanoparticle carriers, for example, can bypass the blood-brain barrier to deliver drugs to the central nervous system, offering new treatment avenues for neurological disorders such as Alzheimer's disease and brain tumors [51]. Similarly, hydrogel-based delivery systems can penetrate the tumor microenvironment to deliver therapeutics to cancer cells, overcoming obstacles such as hypoxia and interstitial fluid pressure.

To address the challenges associated with conventional therapeutic models, there is a growing need for personalized approaches in biomaterials-based hydrogel design and application. Key strategies for advancing personalized hydrogel-based therapies include:

- Integration of patient-specific parameters, such as tissue characteristics, immune response, and genetic factors, into biomaterial selection processes to optimize biocompatibility and therapeutic efficacy [43].

- Development of targeted drug delivery systems that incorporate patient-specific factors, such as tumor biomarkers or disease characteristics, to achieve site-specific drug release and enhance therapeutic outcomes [52].

- Customized fabrication of hydrogel scaffolds with tailored mechanical properties, bioactive cues, and degradation kinetics to promote tissue-specific regeneration and integration [53].

4.1 Importance of drug delivery in medicine

In the realm of modern medicine, drug delivery stands as a cornerstone of therapeutic interventions, facilitating the precise administration of pharmaceutical agents to targeted sites within the body. From alleviating symptoms to curing diseases, drug delivery systems play a pivotal role in optimizing therapeutic outcomes while minimizing adverse effects. Conventional drug delivery systems can have many side effects because they have poor bioavailability, variations in plasma drug levels and do not demonstrate sustained release. Also, some excipients of synthetic origin contained in classic medicines, such as coloring agents, perfumes, antioxidants, anti-caking agents, binding agents, solvents and lubricants, sweeteners, flavoring agents, preservatives or solubilizing agents, can induce nausea, dizziness, headaches, can be metabolized in certain organs, and can promote the onset or worsening of certain diseases over time. For this reason, modern treatment systems based on biomaterials, of natural origin, with improved efficiency and minimal or absent side effects, are at the top of scientific research in the biomedical field.

Drug delivery is a critical component of medical treatment, influencing the efficacy, safety, and patient compliance of pharmaceutical interventions. Several factors underscore the importance of drug delivery in medicine:

i. Drug delivery systems enable the controlled release and targeted delivery of pharmaceutical agents to specific sites within the body, maximizing therapeutic efficacy while minimizing systemic side effects. By optimizing drug concentrations at the site of action, drug delivery systems enhance the effectiveness of treatments for various diseases, ranging from cancer to infectious diseases.

ii. Drug delivery systems offer convenience and ease of administration, enhancing patient compliance with prescribed treatment regimens. Whether in the form of oral tablets, transdermal patches, or injectable formulations, optimized drug delivery systems streamline medication administration, reducing the burden on patients and improving treatment adherence.

iii. Drug delivery systems can be designed to provide sustained release or prolonged action of pharmaceutical agents, maintaining therapeutic drug levels in the body over an extended period. By controlling drug release kinetics, drug delivery systems optimize pharmacokinetics and pharmacodynamics, ensuring continuous and effective treatment while minimizing fluctuations in drug concentrations.

iv. Advances in drug delivery technologies have paved the way for personalized medicine approaches, tailoring treatment regimens to individual patient characteristics and disease profiles. By incorporating patient-specific factors such as genetic polymorphisms, disease biomarkers, and pharmacokinetic parameters, drug delivery systems enable precision medicine strategies that optimize therapeutic outcomes and minimize adverse events.

v. Drug delivery systems can overcome biological barriers such as the blood-brain barrier, gastrointestinal mucosa, and tumor microenvironment, enabling the delivery of therapeutics to otherwise inaccessible sites. By incorporating targeting ligands, nanoparticles, or drug conjugates, drug delivery systems enhance drug penetration and bioavailability, overcoming physiological barriers and improving treatment efficacy for challenging conditions.

vi. Drug delivery systems enable the co-delivery of multiple therapeutic agents, facilitating combination therapies that target multiple pathways or disease processes simultaneously. By encapsulating drugs within the same carrier or incorporating multiple drug-loaded nanoparticles, drug delivery systems enhance synergistic effects, overcome drug resistance, and improve therapeutic outcomes for complex diseases such as cancer and infectious diseases.

4.2 Various drug delivery systems

Drug delivery systems encompass a diverse array of technologies designed to optimize the administration, targeting, and release of pharmaceutical agents within the body, enhancing therapeutic efficacy and minimizing side effects [54].

4.2.1 Conventional drug delivery systems

Oral drug delivery remains one of the most common and convenient methods of medication administration. Oral tablets, capsules, and liquid formulations offer ease of use, patient compliance, and widespread availability. These formulations undergo dissolution and absorption in the gastrointestinal tract, delivering drugs to systemic circulation for distribution to target tissues. Modified-release formulations, such as extended-release tablets and gastro-resistant capsules, enable controlled drug release and prolonged therapeutic action, enhancing treatment efficacy and patient convenience.

Transdermal drug delivery systems deliver drugs through the skin for systemic absorption, bypassing the gastrointestinal tract and first-pass metabolism. Transdermal patches, creams, and gels offer sustained release of drugs over an extended period, providing steady plasma concentrations and minimizing fluctuations in drug levels. These systems are commonly used for delivering hormones, analgesics, and cardiovascular medications, offering advantages such as improved patient compliance, reduced systemic side effects, and avoidance of gastrointestinal irritation.

Inhalation drug delivery systems deliver medications directly to the lungs for rapid absorption and therapeutic effect. Metered-dose inhalers (MDIs), dry powder inhalers (DPIs), and nebulizers are commonly used for delivering bronchodilators, corticosteroids, and other respiratory medications. Inhalation therapy offers targeted delivery to the site of action, rapid onset of action, and reduced systemic exposure, making it an effective approach for managing respiratory diseases such as asthma and chronic obstructive pulmonary disease (COPD).

4.2.2 Advanced drug delivery systems

Nanoparticle-based drug delivery systems utilize nanoscale carriers such as liposomes, polymeric nanoparticles, and lipid nanoparticles to encapsulate and deliver drugs to target tissues. These carriers offer advantages such as enhanced drug solubility, prolonged circulation time, and targeted delivery to specific cells or tissues. Nanoparticle formulations enable the controlled release of drugs, reduce systemic toxicity, and improve therapeutic efficacy, making them promising platforms for cancer therapy, gene delivery, and targeted drug delivery to the brain.

Hydrogel-based drug delivery systems utilize hydrophilic polymer networks to encapsulate drugs and release them in a controlled manner. These systems are used for localized drug delivery to tissues, wound healing, and sustained release of therapeutic agents. Injectable hydrogels offer minimally invasive delivery routes and conformal contact with tissues, making them suitable for tissue engineering, regenerative medicine, and controlled drug release applications.

Implantable drug delivery devices are surgically implanted within the body to deliver drugs over an extended period. These devices include drug-eluting stents, subcutaneous implants, and intravitreal implants, offering localized drug delivery and sustained release of therapeutics. Implantable devices provide advantages such as improved patient compliance, reduced dosing frequency, and enhanced therapeutic efficacy, particularly for chronic conditions requiring long-term treatment.

Targeted drug delivery systems aim to deliver drugs selectively to diseased tissues or cells while minimizing exposure to healthy tissues. These systems utilize targeting ligands, antibodies, or stimuli-responsive materials to achieve site-specific drug delivery. Targeted drug delivery offers advantages such as increased drug accumulation at the target site, reduced systemic toxicity, and enhanced therapeutic efficacy. Examples include antibody-drug conjugates, targeted nanoparticles, and stimuli-responsive hydrogels for cancer therapy, inflammatory diseases, and infectious diseases.

4.3 Advantages of using hydrogels for drug delivery

Hydrogels, with their unique properties and versatile characteristics, have emerged as promising platforms for drug delivery applications. From controlled release to targeted delivery, hydrogels offer numerous advantages that make them attractive for pharmaceutical formulations [55]. Hydrogels offer precise control over drug release kinetics, enabling sustained, controlled release of pharmaceutical agents over extended periods. The porous structure of hydrogels allows drugs to be encapsulated within the polymer matrix, from which they can diffuse or be variably released in response to external stimuli, from zero-order kinetics to pulsatile or stimuli-responsive release.

Hydrogels can be engineered to achieve targeted drug delivery to specific tissues or cells within the body. Functionalization of hydrogels with targeting ligands, antibodies, or peptides enables selective binding to receptors or biomarkers expressed on target cells, enhancing drug accumulation and therapeutic efficacy while minimizing off-target effects. Targeted drug delivery using hydrogels offers advantages such as increased drug bioavailability, reduced systemic toxicity, and enhanced treatment outcomes for diseases such as cancer, inflammation, and infectious diseases.

5. Design and formulation of biomaterial-based hydrogels

5.1 Engineering hydrogel composition

The design of biomaterial-based hydrogels begins with the selection and engineering of hydrogel composition. Natural polymers such as collagen, alginate, hyaluronic acid, and chitosan offer inherent biocompatibility and bioactivity, mimicking the extracellular matrix of native tissues. Synthetic polymers such as polyethylene glycol (PEG), polyacrylamide (PAA), and poly(N-isopropylacrylamide) (PNIPAAm) offer precise control over mechanical properties, degradation kinetics, and responsiveness to stimuli. Hybrid hydrogels, combining elements of natural and synthetic polymers, offer synergistic properties that enhance biocompatibility, mechanical strength, and tunability.

5.2 Crosslinking strategies for hydrogel formation

The formation of hydrogels relies on crosslinking strategies that create stable networks of polymer chains. Various crosslinking methods, including physical, chemical, and biological approaches, offer distinct advantages and limitations for hydrogel synthesis. By selecting appropriate crosslinking strategies, researchers can tailor hydrogel properties such as mechanical strength, swelling behavior, and degradation kinetics to meet specific application requirements.

5.3 Modulating hydrogel properties and functionality

The properties and functionality of hydrogels can be modulated through various strategies, including molecular design, polymer modification, and incorporation of bioactive molecules. By tuning parameters such as polymer concentration, crosslinking density, and network architecture, researchers can control hydrogel stiffness, porosity, and water retention capacity. Stimuli-responsive hydrogels, designed to undergo reversible changes in response to external stimuli such as temperature, pH, or light, offer opportunities for on-demand drug release and dynamic modulation of hydrogel properties.

5.4 Methods for preparing hydrogels

5.4.1 Polymerization techniques

Free radical polymerization: Free radical polymerization is a commonly used method for synthesizing hydrogels from monomers such as acrylic acid, acrylamide, or methacrylate derivatives. Initiators such as ammonium persulfate (APS) and N,N,N′,N′-tetramethylethylenediamine (TEMED) are used to initiate polymerization, leading to the formation of crosslinked polymer networks [56].

Crosslinking Polymerization: Crosslinking polymerization involves the crosslinking of preformed polymer chains to create hydrogel networks. Crosslinking agents such as glutaraldehyde, genipin, or poly(ethylene glycol) diacrylate (PEGDA) are used to link polymer chains via covalent bonds, resulting in hydrogels with tunable mechanical properties and degradation kinetics [57].

5.4.2 Physical gelation methods

Temperature-induced gelation: Temperature-induced gelation relies on the reversible sol-gel transition of polymers in response to changes in temperature. Polymers such as poly(N-isopropylacrylamide) (PNIPAAm) undergo a phase transition near their lower critical solution temperature (LCST), resulting in the formation of hydrogels at elevated temperatures and solubilization at lower temperatures [58].

Ionic gelation: Ionic gelation involves the crosslinking of polymers via ionic interactions between oppositely charged ions. Polymers such as alginate or chitosan form hydrogels in the presence of multivalent cations such as calcium ions (Ca^{2+}), resulting in the formation of stable crosslinked networks [59].

5.4.3 Chemical modification techniques

Functionalization: Hydrogel properties can be modified through chemical functionalization of polymer chains. Functional groups such as carboxylic acids, amino groups, or thiol groups can be introduced onto polymer chains via chemical reactions such as amidation, esterification, or thiolene click chemistry, enabling tailored hydrogel properties and functionalities [60].

Surface modification: Surface modification of hydrogels involves the attachment of bioactive molecules, cell-adhesive peptides, or growth factors onto hydrogel surfaces. Surface modification techniques such as physical adsorption, covalent immobilization, or layer-by-layer assembly enable the creation of bioactive hydrogels with enhanced cell adhesion, proliferation, and differentiation properties.

5.4.4 Solution casting and molding

Solution casting and molding involve the preparation of hydrogel precursor solutions followed by casting into molds and subsequent gelation. This method allows for the fabrication of hydrogel films, discs, or shapes with precise control over size, shape, and thickness.

5.4.5 Electrospinning

Electrospinning is a technique used to fabricate hydrogel fibers with diameters ranging from nanometers to micrometers. Hydrogel precursor solutions are electrospun into fine fibers using an electric field, resulting in the formation of hydrogel mats or scaffolds with high surface area and porosity.

5.4.6 3D printing

3D printing, also known as additive manufacturing, enables the fabrication of complex three-dimensional hydrogel structures layer by layer. Hydrogel precursor solutions are deposited onto a substrate using a computer-controlled nozzle or laser, allowing for the precise spatial patterning of hydrogel materials.

5.5 Characterization techniques

Swelling Ratio Measurement: The swelling ratio of hydrogels, defined as the ratio of swollen weight to dry weight, is often measured to assess hydrogel swelling behavior and water retention capacity.

Mechanical Testing: Mechanical testing techniques such as tensile testing, compression testing, or rheological analysis are used to evaluate the mechanical properties of hydrogels, including stiffness, elasticity, and viscoelastic behavior.

Drug Release Studies: Drug release studies are conducted to assess the release kinetics of therapeutic agents from hydrogel matrices. Techniques such as spectrophotometry, chromatography, or imaging are used to quantify drug release over time.

Biocompatibility Assessment: Biocompatibility assessment involves evaluating the cytotoxicity, cell adhesion, and cell proliferation of hydrogels using *in vitro* cell culture assays or *in vivo* animal studies.

Methods for preparing hydrogels encompass a wide range of synthesis, modification, fabrication, and characterization techniques tailored to achieve specific properties and applications. From polymerization techniques to physical gelation methods, chemical modification strategies, and fabrication techniques, researchers and engineers have at their disposal a versatile toolkit for designing and engineering hydrogels with precise control over structure, properties, and functionality [61].

5.6 Incorporation of therapeutic agents into hydrogel matrices

Hydrogels serve as versatile carriers for the encapsulation and controlled release of therapeutic agents, ranging from small molecules to proteins and nucleic acids [62]. By incorporating therapeutic agents into hydrogel matrices, researchers can achieve precise control over drug release kinetics, improve drug stability, and enhance therapeutic efficacy.

5.6.1 Encapsulation techniques

Physical entrapment involves the encapsulation of therapeutic agents within hydrogel matrices through physical interactions such as hydrogen bonding, electrostatic interactions, or hydrophobic interactions. Therapeutic agents are simply mixed with hydrogel precursor solutions before gelation, resulting in a uniform distribution of drugs throughout the hydrogel matrix. Chemical conjugation entails the covalent attachment of therapeutic agents to polymer chains within hydrogel matrices. Functional groups on therapeutic agents and polymer chains are chemically modified to facilitate covalent bonding, ensuring the stable incorporation of drugs into hydrogel networks.

5.6.2 Key considerations

i. Hydrogels and encapsulated therapeutic agents must be biocompatible to ensure compatibility with biological systems and minimize adverse effects on cells and tissues.

ii. Encapsulated therapeutic agents should remain stable within hydrogel matrices to maintain drug potency and efficacy throughout the release process.

iii. The release kinetics of therapeutic agents from hydrogel matrices should be carefully tuned to achieve desired release profiles, balancing factors such as diffusion rates, degradation rates, and stimuli responsiveness.

5.7 Strategies for controlling drug release from hydrogels

Controlling drug release from hydrogels is crucial for optimizing therapeutic efficacy, minimizing side effects, and achieving desired pharmacokinetic profiles.

5.7.1 Diffusion-controlled release

The structure of hydrogel matrices, including porosity, mesh size, and network density, influences the diffusion of therapeutic agents through the hydrogel network. Tuning these parameters allows for controlled diffusion of drugs and modulation of release kinetics. The concentration of therapeutic agents loaded into hydrogel matrices affects drug release kinetics, with higher loading concentrations typically resulting in faster release rates due to increased concentration gradients.

5.7.2 Degradation-controlled release

Hydrogel degradation kinetics, determined by factors such as polymer composition, crosslinking density, and degradation mechanisms, influence drug release rates. Degradable hydrogels undergo gradual breakdown, releasing encapsulated drugs as the polymer matrix degrades over time. Enzyme-responsive hydrogels incorporate peptide sequences susceptible to enzymatic cleavage within the polymer network. Enzyme activity in the surrounding environment triggers hydrogel degradation and subsequent release of therapeutic agents, enabling spatiotemporal control over drug release.

5.7.3 Stimuli-responsive release

pH-responsive hydrogels undergo structural changes in response to variations in pH, enabling the triggered release of therapeutic agents under acidic or basic conditions. pH-sensitive polymers such as poly(acrylic acid) (PAA) or poly(histidine) (PHis) are commonly used to design pH-responsive hydrogels for targeted drug delivery to specific physiological environments [63]. Temperature-responsive hydrogels exhibit phase transitions in response to changes in temperature, resulting in sol–gel transitions and modulation of drug release kinetics. Thermoresponsive polymers such as poly(N-isopropylacrylamide) (PNIPAAm) undergo reversible phase transitions near their lower critical solution temperature (LCST), enabling the temperature-triggered release of therapeutic agents [64].

5.7.4 External stimuli-responsive release

Light-responsive hydrogels incorporate photoreactive moieties that undergo conformational changes or crosslinking upon exposure to specific wavelengths of light. Light-triggered hydrogel systems enable spatially controlled release of therapeutic agents using light patterning techniques such as photolithography or laser irradiation. Magnetic-responsive hydrogels incorporate magnetic nanoparticles within the polymer matrix, enabling remote control over drug release using external magnetic fields. Magnetic stimuli induce mechanical deformation or swelling of hydrogel networks, facilitating the on-demand release of therapeutic agents at target sites.

6. Biomaterials-based hydrogels in skin tissue engineering, surgical defects, and wound healing

6.1 Role of hydrogels in wound management

The outer layer of the human body, the skin, is our largest organ, and the first defense barrier against harmful environmental influences [65, 66]. It is incredibly versatile and performs many vital functions. The skin has the ability to regenerate itself naturally, but this self-healing ability can be diminished for wounds larger than about 4 centimeters in diameter [67]. *Bandages* are applied to help the wound healing process and also to protect the wounds from various infections. Currently, a variety of modern products have been commercially launched for wound care, including *foams*, *films*, *hydrocolloids*, *hydrogels*, and *hydrofibers*. Hydrogels are ideal wound dressings due to their exceptional biological properties [68–71]. Dressings based on hydrogels are considered advanced successful systems and particularly useful tools in biomedical practice.

Hydrogels can be designed with a wide variety of characteristics to obtain *biocompatible*, *stable,* and *bioresorbable* polymeric matrices for applied biomedicine such as skin tissue engineering and wound management [72, 73]. A crucial attribute of the use of hydrogels in wound healing is to ensure *facile application*, *long-lasting adhesiveness,* and the *ability to be easily removed* from human skin without causing any damage or leaving traces [74–76]. In addition, they are considered therapies with great potential to reduce scar formation in skin wounds due to their high water content. As dressing materials, they can be used both for exudative wounds and for dry wounds. The main constraints of medical adhesive hydrogels are the strength to adhere to soft tissue surfaces and the resistance to cyclic stresses in the moist and dynamic environments surrounding the tissues [77]. Hydrogels do not cause irritation, allow the passage of metabolites, and do not interact with biological tissues [78]. Hydrogels are used to transport various drugs to areas affected by wounds, benefiting from their porous, hydrophilic structure [2]. A promising strategy to avoid scar formation following skin injury involves employing a hydrogel as a pro-regenerative matrix combined with growth factors and various cell types to stimulate tissue regeneration. Its adhesive attributes allow it to adhere to hard and soft tissues, being a valuable material in surgery, orthopedics, dentistry, ophthalmology, technical-sanitary instrumentation, or surgical tools [79–82].

Polysaccharide-based hydrogels used as dressings for wound healing ensure several simultaneous activities to support complete recovery and have the following role [83]:

1. **Antioxidant**. They lead to the reduction of the level of reactive oxygen species that affect the antioxidant capacity necessary to promote the healing of diabetic wounds. Oxidative stress inhibits healing progress in many categories of skin wounds, such as bacterial infections, erosions, scars, ulcers, burns, or acute trauma.

2. **Anti-inflammatory**. Inhibition of anti-inflammatory responses is a key factor in wound healing, especially chronic wounds. Incorporating drugs with anti-inflammatory properties into the bioinspired hydrogel matrix is an advantageous approach that helps increase the anti-inflammatory effects.

3. **Antibacterial**. The incorporation of biomaterials into wound dressings increases their ability to inhibit the growth of a wide spectrum of bacteria and reduces the risk of infection, avoiding dangers such as necrosis, sepsis, or other fatal hazards.

4. **Hemostatic**. Hemostasis is the first and very important stage of wound healing because an uncontrolled hemorrhage can lead to death. Fast hemostasis can be achieved with dressings based on hydrogels with a porous structure, which are

advantageous because they have excellent capacities for absorbing large amounts of injury exudate and controlling wound bleeding.

5. **Supports the growth of skin cells and granulation tissues.** Hydrogels are used as carriers for natural compounds or drugs for skin healing. Embedding the drugs or various slow-release bioactive constituents into multifunctional hydrogel scaffolds can lead to an increase in the long-term healing capacity of wounds.

6.2 Delivery of growth factors and cells for tissue skin regeneration

Wound healing is a physiological process that occurs continuously, in several stages: *hemostasis*, *inflammation*, *proliferation*, or formation of granulation tissue, *remodeling* or maturation of the newly formed tissue [84]. It requires direct dynamic interactions between the extracellular matrix (ECM) and growth factors or indirectly, for example, the binding of different types of cells to the ECM [85, 86]. Growth factors are polypeptide molecules secreted by cells, with a signaling role that adjusts cellular responses corresponding to the stages of healing. During the healing of skin wounds, various growth factors are secreted and released that help synthesize collagen and regenerate the epidermis [87]. The benefits of the administration of growth factors are limited because they have low stability *in vivo*, slow absorption through the skin, and can be eliminated with the exudate. In clinical practice, the efficient and safe delivery of growth factors requires controlled delivery systems. However, the action time of the growth factors can be increased if they are loaded in natural hydrogel matrices [88]. They have many functional groups that provide suitable binding sites to form stable bonds with growth factors [89].

A promising strategy to regulate skin healing and avoid scar formation after skin injuries involve the use of hydrogel as a pro-regenerative matrix combined with growth factors and different cell types to act directly on local wounds and to stimulate tissue regeneration.

6.3 Clinical applications and case studies

Numerous studies have presented the results of research with the perspective of developing materials for healing wounds with adhesive and hemostatic properties. The research focused on composite hydrogels based on nanoparticles inspired by mussels, recognized for their exceptional ability to adhere in an aqueous environment and for rapid hardening due to the contained Byssus proteins [90]. A very recent study reported the obtaining of a therapeutic and bioinspired hydrogel based on crosslinked CS/COL combinatorial biomacromolecules. Both dry and wet hydrogels obtained incorporated green synthesized AgNPs loaded with cefotaxime sodium. They showed good antimicrobial activity against gram-positive and negative bacteria. Wound healing activity performed on injured rats demonstrated complete wound closure after 2 and 3 weeks, respectively, fully restoring skin function [91]. A novel pharmaceutical hydrogel containing bacteriomimetic microparticles based on membrane vesicles (MV) produced by Lactobacilli was designed for wound healing. The anti-inflammatory effect of the hydrogel was tested on primary human peripheral blood mononuclear cells, and scar formation was observed in a mouse model *in vivo*. The hydrogel showed anti-inflammatory effects *in vitro* and a good ability to heal wounds and reduce scar formation *in vivo* [92].

The fabrication of bioinspired dressings for wounds is of significant interest from the perspective of the increasing incidence of chronic diseases and the increase of resistance to antibiotics. The researchers approached this topic of interest and created hydrogels based on hyaluronic acid methacrylate and gelatin methacrylate, incorporating selenium nanoparticles. The nanoparticles were obtained through a nontoxic microwave-assisted hydrothermal synthesis strategy. The effectiveness of nanoparticle-loaded hydrogels in wound healing was evaluated by the *in vitro* scratch test. Experiments have shown that nanoparticles significantly improve results compared to other types of nanoparticles, and hydrogels can act as effective dressings to facilitate wound healing [93].

It is known that cardiovascular diseases cause many millions of deaths worldwide every year, and treatments after surgical interventions require the development of new and more effective therapies and medical technology. It is an important direction for top research, and numerous biomaterials-based approaches have extensively investigated this critical area [82]. Open heart operations or heart transplants are major and high-risk interventions, including the risk of infection. Various research has addressed obtaining cardiac patches that provide temporary support to the infarcted area for the administration of cells or bioactive factors or anti-inflammatory drugs. Researchers have developed a novel hydrogel patch for sustained release directly into the infarcted heart to reduce inflammation. The adhesive hydrogel that can be painted was obtained based on dextran-aldehyde (dex-ald) and gelatin, incorporating the anti-inflammatory protein, ANGPTL4. Experiments performed on cardiac tissue treated with ANGPTL4-loaded hydrogel patches showed increased vascularity, reduced inflammatory macrophages, and structural maturation of cardiac cells. In addition, the developed hydrogel showed suitable tissue adhesion, excellent mechanical stability, and sustained release of anti-inflammatory drugs. The results of the *in vivo* experiments encourage the authors to consider the hydrogel patch a useful tool for the repair of various tissues, including the heart, muscles, and cartilage [94].

The development of *biotherapeutic hydrogels* offers new treatment options for a wide range of conditions, including *acute wounds*, *diabetic wounds*, and *burns*. The healing process can be affected by many factors, such as bacterial infections, high levels of reactive oxygen species, macrophage dysfunction, sustained hypoxia, excess pro-inflammatory cytokines, age, sex, obesity, diabetes, nutrition, or medication. All these elements are taken into consideration when designing new syntheses of *smart hydrogels* to provide *multifunctional abilities* that allow them to *effectively respond* and *accelerate wound healing*. Likewise, they can have different *properties in real time*, such as *reactivity to stimuli*, *injectable self-healing*, *shape memory,* and *conductive monitoring* [95–102].

7. Hydrogels in tissue engineering

7.1 Use of hydrogels as scaffolds for tissue regeneration

The potential to coexist and interconnect within specific physiological systems or around tissues without causing damage is the key factor that distinguishes biomaterials from other types of materials. In the modern innovative era, advanced techniques of combining biomimetic materials and incorporating cells and bioactive molecules design tissue engineering scaffolds with distinct 3D structure that ensure mechanical aid for cells in engineered tissues and simulate the native extracellular matrix [1]. The manufacture of biomaterial-based scaffolds that do not induce immune reactions has a decisive contribution in stimulating angiogenic and osteogenic progressions and is based on the selection of the most suitable biomaterial. In the area of tissue engineering, hydrogels based on biomaterials are effective in promoting the healing of bone defects. However, inadequate vascularization in large bone defects is a huge challenge for clinical bioengineering and limits progress in the production of bone substitutes [103].

The biodegradability of the scaffold should be controllable and tunable to allow useful remodeling. This remodeling particularizes vascularization, cellular differentiation, and degradation of the scaffold, so that at the end, this scaffold is substituted by the target tissue [104]. Osteoarthritis is a degenerative disease whose major pathological features are articular cartilage defects, which in turn amplify inflammation in the joint. Cartilage is a significant tissue whose damage can amplify the deterioration of joint function. The tissue's limited self-repair capacity is insufficient and significant regeneration will not occur due to the complex structure of cartilage, where there are no blood vessels, nerves, or lymphatic tissue. Biomaterials-based hydrogels with elastic structures, with smooth surfaces and a high-water content can be designed with properties adapted for the repair of different types of cartilage defects. Numerous studies have developed and applied advanced hydrogels *in vitro* or *in vivo* created for the needs of cartilage tissue engineering and precision medicine.

A very recent study presented the development of a semi-flexible hydrogel that mimics the stiffness of natural tissue and facilitates bone regeneration. The hydrogel based on elastic fibers of reticulated fibrinogen and collagen consumes energy and makes the transition from soft to hard, changing its internal state along with body temperature. Due to its hydrophilicity, the hydrogel quickly adhered to the surface, then became rigid, reducing inflammation in the early stages and contributing to the formation of bone tissue. Due to the ability of accelerated regeneration, this bioinspired hydrogel has the potential to be applied to various other tissues [105].

A new formulation of injectable hydrogel based on kappa-carrageenan-co-N-isopropyl acrylamide (κC-*co*-NIPAAM) was made by free radical polymerization and antisolvent evaporation technique. The results of digital X-ray investigation using an *in vivo* bone defect model showed that the synthesized hydrogel improved bone regeneration [106].

Hydrogels serve as supporting matrices to deliver cardiomyocytes and stem cells in regenerating cardiac tissue into infarcted cardiac muscle. Among the hydrogel types used for cardiac tissue are natural polymer hydrogels, synthetic polymer hydrogels, or hybrid hydrogels [107]. Research highlights the hydrogels with elastomeric, conductive, and oxygen-releasing properties and stimuli-responsive hydrogels, which have the ability to react to a range of physical, chemical, and biological stimuli, mimicking the cardiac tissue [108, 109]. Liver tissue engineering projects the development of physiologically applicable liver models. The recent *in vitro* models of livers made with the help of bioengineering are encouraging for testing drugs, toxicological studies or as disease models and as a possible alternative, in the future, to insufficient donor organs, to treat end-stage liver diseases [110].

7.2 Encapsulation of cells within hydrogel matrices

It is now known that the synergistic interaction between immune cells and cellular encapsulation is responsible for a proper regenerative process, which is the basis of the new concept of modular tissue engineering. Thus, cell-laden hydrogels are being investigated as native-like systems for various applications in regenerative medicine and hard tissue repair. The development of *cell encapsulation strategies* in different biodegradable hydrogel formulas provides several advantageous features for tissue engineering applications, such as (i) ease of application, (ii) a highly hydrated substrate that provides a favorable environment for cell and tissue growth, (iii) the possibility of it is formed *in vivo*, and (iv) controlled degradation [111].

Degradation regulation can be achieved by (i) the selection and use of natural biopolymers that are susceptible to enzymatic degradation and (ii) the integration into the hydrogel of segments that are less stable from a hydrolytic or enzymatic point of view. Since the encapsulation of the cells occurs together with the gelation process, the number of suitable formulations is limited.

7.3 Challenges and advancements in tissue engineering with hydrogels

One of the current scientific challenges is the study of the structure of hydrogels derived from the extracellular matrix proteins of natural tissues, the quantification of their composition and their extensive characterization, for their application as injectable or preformed cell delivery matrices in tissue engineering and regenerative medicine.

Protein-based biopolymers obtained from natural tissues have a hierarchical configuration in their native state. They are isolated from their natural tissue through various advanced processes and solubilized in an aqueous solution to be remodeled into injectable or preformed hydrogels for tissue engineering and regenerative applications.

Bio gels are advanced materials based on novel biopolymers derived from proteins, lipids, nucleic acids, and carbohydrates, which provide the bioactive amino acid sequences required for the adhesion, growth, and maturation of encapsulated cells.

8. Hydrogels in ophthalmic and dental applications

The recent applications of hydrogels in ophthalmology highlight their capacity as versatile, efficient, biocompatible, and adaptable therapeutic tools. They prove the ability to target, control, and sustain drug release to the posterior segment of the eye, minimizing invasive operations and improving patient results [112]. Also, hydrogels are useful in postoperative drug delivery and disease detection. However, most hydrogel-based studies remain in preclinical stages, requiring rigorous clinical evaluations [113].

Hydrogels are useful in ocular applications such as *soft contact lenses* (SCLs) or as materials during contact lens development and can be integrated with nature-inspired drug immobilization and release strategies, providing useful tools for ocular delivery systems of medicines [114].

Oral tissue engineering uses dental biomaterials that require a dual purpose: combating bacterial infections and promoting tissue growth. Biomaterial-based hydrogels demonstrate exceptional potential for oral tissue regeneration and drug delivery [115].

Materials based on hydrogels must face specific challenges, such as the oral environment with sudden temperature changes, the presence of different categories of bacteria, and pH fluctuations caused by saliva and biofilms [116]. Controlled syntheses of hydrogels can regulate their porous structure to achieve pores of the right size and shape with the surrounding tissues, promoting cell adhesion and growth and being useful in regenerative therapy [117, 118].

Among the *advantages of hydrogels for oral applications*:

- high fluid absorption capacity,
- antibacterial properties,
- effective drug release carriers,
- response to external physical, chemical, and biological stimuli,
- promotes the regeneration of oral tissue.

9. Targeted therapy in theragnostic

Hydrogels are promising tools widely applied in the treatment of cancer by chemotherapy, radiotherapy, immunotherapy, hyperthermia, photodynamic therapy, and photothermal therapy (**Figure 2**) [119].

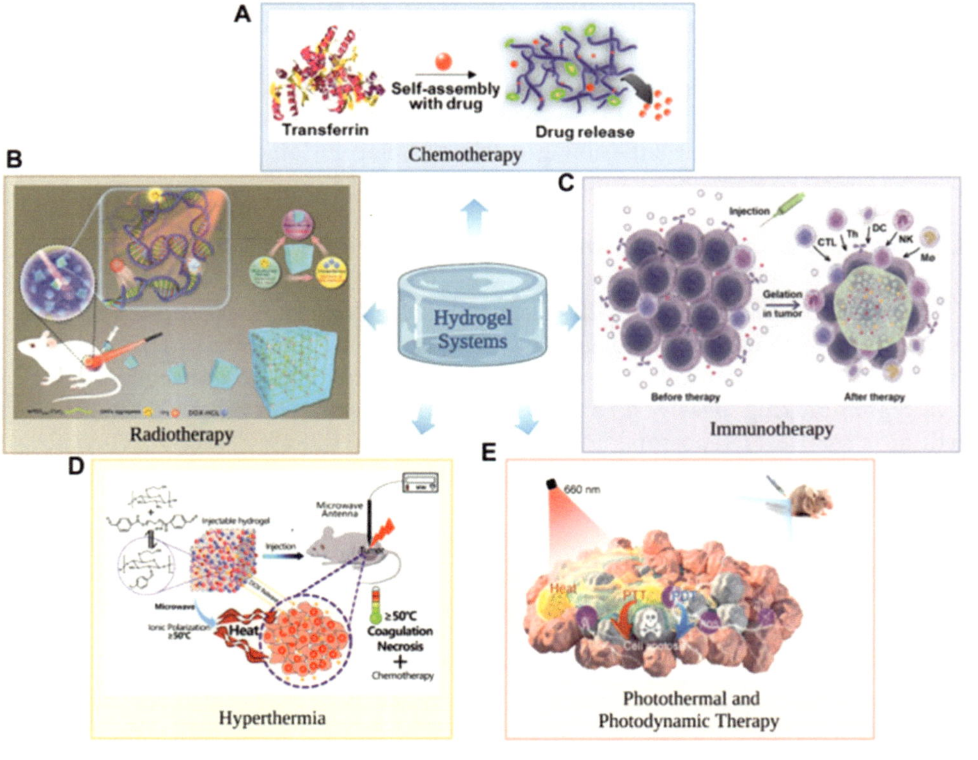

Figure 2.
The applications of hydrogels (A) cancer chemotherapy; (B) radiotherapy; (C) immunotherapy; (D) hyperthermia; (E) photodynamic therapy and photothermal therapy [119].

DOI: http://dx.doi.org/10.5772/intechopen.1004826

The *main advantages of hydrogels as excellent carriers of new drugs in cancer treatment* are:

- their use as controlled and precise systems for the continuous and sequential release of drugs, chemotherapeutics, radionuclides, immunosuppressants, hyperthermia agents, phototherapy agents, and other substances.

- targeting different sites and categories of cancer, thus reducing the dose of common drugs, and improving the efficiency of the treatment.

- providing intelligent responses to environmental changes depending on internal and external environmental stimuli, with the possibility of remote control and the release on demand of various active anticancer substances.

- the combination of imaging and therapeutic applications in an only one theragnostic platform by incorporating different contrast agents or nanoparticles in the hydrogel matrix and monitoring in real time the response to the treatment and the stage of the disease.

- the possibility of adjusting the treatment parameters in real time, optimizing the therapeutic results, and minimizing the development of drug resistance and side effects.

As targeted therapy in theragnostic, injectable hydrogels have a huge potential due to the possibility of *in vivo* application and their distinctive way of administration in the human body. Design strategies aim to develop multifunctional injectable hydrogels with good adhesion, elasticity, and bioresorbability *in vivo* to be applied via a minimal route. In the structure of the hydrogel, functional bionanomaterials can be incorporated to be used in various diagnostic and therapeutic applications *in vivo*, including on representative organs such as the skin, liver, heart, and urinary bladder [120].

10. Challenges and future perspectives

Several challenges persist in the field, which researchers are actively addressing to advance the field.

1. **Biocompatibility and immunogenicity**
Ensuring the biocompatibility of hydrogels is essential for their safe and effective use in biomedical applications. Hydrogel degradation products or residual crosslinking agents may induce inflammatory responses or immune reactions *in vivo*, limiting their clinical utility.

2. **Mechanical properties and stability**
Achieving adequate mechanical properties and stability is critical for the structural integrity and functional performance of hydrogels *in vivo*. Weak mechanical properties or rapid degradation may compromise the longevity and efficacy of hydrogel-based therapies.

3. **Control over drug release kinetics**
Precise control over drug release kinetics is essential for optimizing therapeutic efficacy and minimizing side effects in drug delivery applications, particularly in complex biological environments.

4. **Biomimetic tissue engineering constructs**
Fabricating biomimetic tissue engineering scaffolds with appropriate structural and biochemical cues to promote tissue regeneration remains a formidable challenge. Mimicking the complex hierarchical organization and microenvironment of native tissues in hydrogel-based constructs is essential for successful tissue engineering outcomes.

5. **Clinical translation and regulatory hurdles**
Transitioning hydrogel-based technologies from the laboratory to clinical practice poses significant time, regulatory, and translational challenges. Establishing robust preclinical evaluation protocols, conducting well-designed clinical trials, and addressing manufacturing challenges are critical steps toward successful commercialization and widespread adoption.

6. **Therapeutic efficacy and clinical outcomes**
Demonstrating the therapeutic efficacy and clinical benefits of hydrogel-based therapies in human patients remains a fundamental challenge. Robust preclinical studies and clinical trials are needed to validate the safety, efficacy, and long-term outcomes of hydrogel-based interventions. Rigorous preclinical evaluation and clinical validation of hydrogel-based therapies are essential for establishing their clinical utility and market acceptance.

10.1 Future trends and potential advancements – emerging technologies in hydrogel research

As researchers continue to push the boundaries of hydrogel technology, several emerging trends and potential advancements are shaping the future landscape of hydrogel research, promising to revolutionize for revolutionizing healthcare and biomedical applications.

10.1.1 Smart hydrogels for precision therapeutics

Smart hydrogels, engineered to respond to specific stimuli such as temperature, pH, light, or biomolecular signals, offer unprecedented control over drug release kinetics and therapeutic delivery. These stimuli-responsive hydrogels enable precise spatiotemporal modulation of drug release tailored to the dynamic physiological microenvironment. From on-demand drug release to triggered responses to disease biomarkers, smart hydrogels offer versatile platforms for precision medicine applications.

10.1.2 3D bioprinting for biomimetic tissue engineering

3D bioprinting technology enables the fabrication of complex, biomimetic tissue engineering constructs with precise spatial control over cell distribution, scaffold architecture, and biochemical cues. By integrating cells, biomaterials, and growth factors into customizable 3D structures, bioprinting holds promise for engineering functional tissues and organs *ex vivo*, and organ transplantation.

10.1.3 Bioactive hydrogels for regenerative medicine

Bioactive hydrogels, incorporating bioactive molecules, cell-adhesive peptides, and growth factors, mimic the biochemical cues of the native extracellular matrix to promote tissue regeneration and repair. These bioactive cues facilitate cell adhesion, proliferation, and differentiation within hydrogel scaffolds, fostering tissue-specific regeneration. From promoting bone regeneration and cartilage repair to facilitating cardiac tissue engineering and neural regeneration, bioactive hydrogels offer therapeutic interventions for a wide range of degenerative diseases and traumatic injuries.

10.1.4 Nanotechnology-enabled drug delivery systems

Nanotechnology-based drug delivery systems, incorporating nanoparticles, liposomes, or nanofibers within hydrogel matrices, offer precise control over drug loading, release kinetics, and targeting specificity. These nanocomposite hydrogels enable efficient encapsulation and delivery of therapeutic agents, overcoming biological barriers and enhancing therapeutic efficacy.

10.1.5 Multifunctional hydrogels for theragnostic

Multifunctional hydrogels, integrating therapeutic and diagnostic functionalities within a single platform, enable theranostic applications for disease diagnosis, monitoring, and treatment. These theranostic hydrogels combine drug delivery capabilities with imaging modalities, biosensing capabilities, or controlled release mechanisms, offering integrated solutions for precision medicine.

The future of hydrogel research is characterized by transformative advancements and innovative technologies that promise to revolutionize healthcare and biomedical applications. From smart hydrogels for precision therapeutics to 3D bioprinting for biomimetic tissue engineering, bioactive hydrogels for regenerative medicine, nanotechnology-enabled drug delivery systems, and multifunctional hydrogels for theragnostic, the potential for hydrogel-based technologies is vast and far-reaching.

11. Conclusions

Biomaterials-based hydrogels represent a frontier in biomedical engineering, offering versatile platforms for a wide range of therapeutic applications, including drug delivery, tissue engineering, regenerative medicine, and diagnostics. The unique properties of hydrogels, such as biocompatibility, tunable mechanical properties, and controlled release kinetics, make them invaluable tools for addressing complex biomedical challenges and improving patient outcomes. From the design and formulation of hydrogels to the incorporation of therapeutic agents, control over drug release kinetics, and future directions in research and development, biomaterials-based hydrogels offer transformative opportunities for precision medicine and personalized therapy. Hydrogels have demonstrated remarkable potential in targeted drug delivery, enabling localized delivery of therapeutic agents to specific tissues or organs while minimizing systemic toxicity. By engineering stimuli-responsive hydrogels, researchers can achieve precise control over drug release kinetics, enabling on-demand and site-specific delivery tailored to the dynamic physiological microenvironment. Moreover, hydrogels hold promise for tissue engineering and regenerative medicine applications, where they serve as scaffolds for cell encapsulation, proliferation, and differentiation. Bioactive hydrogels, incorporating growth factors, cell-adhesive peptides, and extracellular matrix components, facilitate tissue-specific regeneration and repair, offering therapeutic interventions for degenerative diseases and traumatic injuries. Looking ahead, future advancements in biomaterials-based hydrogel research are expected to focus on smart hydrogels for precision therapeutics, 3D bioprinting for biomimetic tissue engineering, nanotechnology-enabled drug delivery systems, and multifunctional hydrogels for theragnostic. By harnessing the collective expertise of multidisciplinary researchers, engineers, clinicians, and industry partners, we can anticipate continued innovation and translation of hydrogel-based technologies into clinical practice, ultimately improving patient care and quality of life. In conclusion, biomaterials-based hydrogels hold immense promise for therapeutic applications, offering innovative solutions for addressing unmet medical needs and advancing healthcare. As research in hydrogel science and engineering continues to evolve, the potential for biomaterials-based hydrogels to revolutionize medicine and transform the treatment landscape is boundless.

Conflict of interest

The authors declare no conflict of interest.

Author details

Mariana Chelu* and Adina Magdalena Musuc
Ilie Murgulescu Institute of Physical Chemistry, Bucharest, Romania

*Address all correspondence to: mchelu@icf.ro

References

[1] Mantha S, Pillai S, Khayambashi P, Upadhyay A, Zhang Y, Tao O, et al. Smart hydrogels in tissue engineering and regenerative medicine. Materials (Basel). 2019;**12**(20):3323. DOI: 10.3390/ma12203323

[2] Thang NH, Chien TB, Cuong DX. Polymer-based hydrogels applied in drug delivery: An overview. Gels. 2023;**9**:523. DOI: 10.3390/gels9070523

[3] Jacob S, Nair AB, Shah J, Sreeharsha N, Gupta S, Shinu P. Emerging role of hydrogels in drug delivery systems, tissue engineering and wound management. Pharmaceutics. 2021;**13**(3):357. DOI: 10.3390/pharmaceutics13030357

[4] Ho T-C, Chang C-C, Chan H-P, Chung T-W, Shu C-W, Chuang K-P, et al. Hydrogels: Properties and applications in biomedicine. Molecules. 2022;**27**:2902. DOI: 10.3390/molecules27092902

[5] Chelu M, Musuc AM. Advanced biomedical applications of multifunctional natural and synthetic biomaterials. Processes. 2023;**11**:2696. DOI: 10.3390/pr11092696

[6] Alshangiti DM, El-Damhougy TK, Zaher A, Madani M, Mohamady GM. Revolutionizingbiomedicine:Advancements, applications, and prospects of nanocomposite macromolecular carbohydrate-based hydrogel biomaterials: A review. RSC Advances. 2023;**13**(50):35251-35291. DOI: 10.1039/d3ra07391b

[7] Agrawal A, Hussain CM. 3D-printed hydrogel for diverse applications: A review. Gels. 2023;**9**:960. DOI: 10.3390/gels9120960

[8] Al-Shalawi FD, Mohamed Ariff AH, Jung D-W, Mohd Ariffin MKA, Seng Kim CL, Brabazon D, et al. Biomaterials as implants in the orthopedic field for regenerative medicine: Metal versus synthetic polymers. Polymers. 2023;**15**:2601. DOI: 10.3390/polym15122601

[9] Davis R, Singh A, Jackson MJ, Coelho RT, Prakash D, Charalambous CP, et al. A comprehensive review on metallic implant biomaterials and their subtractive manufacturing. International Journal of Advanced Manufacturing Technology. 2022;**120**(3-4):1473-1530. DOI: 10.1007/s00170-022-08770-8

[10] Spałek J, Ociepa P, Deptuła P, Piktel E, Daniluk T, Król G, et al. Biocompatible materials in otorhinolaryngology and their antibacterial properties. International Journal of Molecular Sciences. 2022;**23**(5):2575. DOI: 10.3390/ijms23052575

[11] Zielińska A, Karczewski J, Eder P, Kolanowski T, Szalata M, Wielgus K, et al. Scaffolds for drug delivery and tissue engineering: The role of genetics. Journal of Controlled Release. 2023;**359**:207-223. DOI: 10.1016/j.jconrel.2023.05.042

[12] Flores-Rojas GG, Gómez-Lazaro B, López-Saucedo F, Vera-Graziano R, Bucio E, Mendizábal E. Electrospun scaffolds for tissue engineering: A review. Macromolecules. 2023;**3**:524-553. DOI: 10.3390/macromol3030031

[13] Trucillo P. Biomaterials for drug delivery and human applications. Materials (Basel). 2024;**17**(2):456. DOI: 10.3390/ma17020456

[14] Ramesh M, Janani R, Deepa C, Rajeshkumar L. Nanotechnology-enabled

biosensors: A review of fundamentals, design principles, materials, and applications. Biosensors (Basel). 2022;**13**(1):40. DOI: 10.3390/bios13010040

[15] Zhang YS, Yao J. Imaging biomaterial-tissue interactions. Trends in Biotechnology. 2018;**36**(4):403-414. DOI: 10.1016/j.tibtech.2017.09.004

[16] Ghimire S, Sarkar P, Rigby K, Maan A, Mukherjee S, Crawford KE, et al. Polymeric materials for hemostatic wound healing. Pharmaceutics. 2021;**13**(12):2127. DOI: 10.3390/pharmaceutics13122127

[17] Gheorghiță D, Moldovan H, Robu A, Bița AI, Grosu E, Antoniac A, et al. Chitosan-based biomaterials for hemostatic applications: A review of recent advances. International Journal of Molecular Sciences. 2023;**24**(13):10540. DOI: 10.3390/ijms241310540

[18] Kim H, Kwak G, Kim K, Yoon HY, Kwon IC. Theranostic designs of biomaterials for precision medicine in cancer therapy. Biomaterials. 2019;**213**:119207. DOI: 10.1016/j.biomaterials.2019.05.018

[19] Alghamdi MA, Fallica AN, Virzì N, Kesharwani P, Pittalà V, Greish K. The promise of nanotechnology in personalized medicine. Journal of Personalized Medicine. 2022;**12**:673. DOI: 10.3390/jpm12050673

[20] Zhu G, Zhang T, Chen M, Yao K, Huang X, Zhang B, et al. Bone physiological microenvironment and healing mechanism: Basis for future bone-tissue engineering scaffolds. Bioactive Materials. 2021;**6**(11):4110-4140. DOI: 10.1016/j.bioactmat.2021.03.043

[21] Kharbikar BN, Zhong JX, Cuylear DL, Perez CA, Desai TA. Theranostic biomaterials for tissue engineering. Current Opinion in Biomedical Engineering. 2021;**19**:100299. DOI: 10.1016/j.cobme.2021.100299

[22] Al-Zyoud W, Haddadin D, Hasan SA, Jaradat H, Kanoun O. Biocompatibility testing for implants: A novel tool for selection and characterization. Materials (Basel). 2023;**16**(21):6881. DOI: 10.3390/ma16216881

[23] Wang Y, Malcolm DW, Benoit DW. Controlled and sustained delivery of siRNA/NPs from hydrogels expedites bone fracture healing. Biomaterials. 2017;**139**:127-138

[24] Rizzi SC, Ehrbar M, Halstenberg S, Raeber GP, Schmoekel HG, Hagenmüller H, et al. Recombinant protein-co-PEG networks as cell-adhesive and proteolytically degradable hydrogel matrixes. Part II: Biofunctional characteristics. Biomacromolecules. 2006;**7**:3019-3029

[25] Filippi M, Born G, Chaaban M, Scherberich A. Natural polymeric scaffolds in bone regeneration. Frontiers in Bioengineering and Biotechnology. 2020;**8**:474. DOI: 10.3389/fbioe.2020.00474

[26] Puertas-Bartolomé M, Mora-Boza A, García-Fernández L. Emerging biofabrication techniques: A review on natural polymers for biomedical applications. Polymers (Basel). 2021;**13**(8):1209. DOI: 10.3390/polym13081209

[27] Marin E, Lanzutti A. Biomedical applications of titanium alloys: A comprehensive review. Materials. 2024;**17**:114. DOI: 10.3390/ma17010114

[28] Amirtharaj Mosas KK, Chandrasekar AR, Dasan A, Pakseresht A, Galusek D. Recent

advancements in materials and coatings for biomedical implants. Gels. 2022;**8**(5):323. DOI: 10.3390/gels8050323

[29] Baino F, Novajra G, Vitale-Brovarone C. Bioceramics and scaffolds: A winning combination for tissue engineering. Frontiers in Bioengineering and Biotechnology. 2015;**3**:202. DOI: 10.3389/fbioe.2015.00202

[30] Krishani M, Shin WY, Suhaimi H, Sambudi NS. Development of scaffolds from bio-based natural materials for tissue regeneration applications: A review. Gels. 2023;**9**(2):100. DOI: 10.3390/gels9020100

[31] Biglari N, Nazarzadeh Zare E. Conjugated polymer-based composite scaffolds for tissue engineering and regenerative medicine. Alexandria Engineering Journal. 2024;**87**:277-299. DOI: 10.1016/j.aej.2023.12.041

[32] Davoodi E, Montazerian H, Mirhakimi AS, Zhianmanesh M, Ibhadode O, Shahabad SI, et al. Additively manufactured metallic biomaterials. Bioactive Materials. 2021;**15**:214-249. DOI: 10.1016/j.bioactmat.2021.12.027

[33] Chelu M, Musuc AM. Polymer gels: Classification and recent developments in biomedical applications. Gels. 2023;**9**:161. DOI: 10.3390/gels9020161

[34] Chelu M, Musuc AM, Popa M, Calderon Moreno J. Aloe vera-based hydrogels for wound healing: Properties and therapeutic effects. Gels. 2023;**9**:539. DOI: 10.3390/gels9070539

[35] Antoine EE, Vlachos PP, Rylander MN. Review of collagen I hydrogels for bioengineered tissue microenvironments: Characterization of mechanics, structure, and transport. Tissue Engineering. Part B, Reviews. 2014;**20**(6):683-696. DOI: 10.1089/ten.TEB.2014.0086

[36] Sun J, Tan H. Alginate-based biomaterials for regenerative medicine applications. Materials (Basel). 2013;**6**(4):1285-1309. DOI: 10.3390/ma6041285

[37] Chelu M, Moreno JC, Atkinson I, Cusu JP, Rusu A, Bratan V, et al. Green synthesis of bioinspired chitosan-ZnO-based polysaccharide gums hydrogels with propolis extract as novel functional natural biomaterials. International Journal of Biological Macromolecules. 2022;**211**:410-424

[38] Zhu J. Bioactive modification of poly(ethylene glycol) hydrogels for tissue engineering. Biomaterials. 2010;**31**(17):4639-4656. DOI: 10.1016/j.biomaterials.2010.02.044

[39] Ansari MJ, Rajendran RR, Mohanto S, Agarwal U, Panda K, Dhotre K, et al. Poly(*N*-isopropylacrylamide)-based hydrogels for biomedical applications: A review of the state-of-the-art. Gels. 2022;**8**(7):454. DOI: 10.3390/gels8070454

[40] Ali F, Khan I, Chen J, Akhtar K, Bakhsh EM, Khan SB. Emerging fabrication strategies of hydrogels and its applications. Gels. 2022;**8**(4):205. DOI: 10.3390/gels8040205

[41] Del Valle LJ, Díaz A, Puiggalí J. Hydrogels for biomedical applications: Cellulose, chitosan, and protein/peptide derivatives. Gels. 2017;**3**:27. DOI: 10.3390/gels3030027

[42] Angaria N, Saini S, Hussain SM, Sharma S, Singh G, Khurana N, et al. Natural polymer-based hydrogels: Versatile biomaterials for biomedical applications. International Journal of Polymeric Materials and

Polymeric Biomaterials. 2024:1-19. DOI: 10.1080/00914037.2023.2301645

[43] Edri R, Gal I, Noor N, Harel T, Fleischer S, Adadi N, et al. Personalized hydrogels for engineering diverse fully autologous tissue implants. Advanced Materials. 2019;**31**(1):e1803895. DOI: 10.1002/adma.201803895

[44] Collins M, N, Cagney L, Thanusha AV. Chapter 6 - hydrogel functionalization and crosslinking strategies for biomedical applications. In: Oliveira JM, Silva-Correia J, Reis RL, editors. Hydrogels for Tissue Engineering and Regenerative Medicine. Cambridge, MA, USA: Academic Press; 2024. pp. 105-137. DOI: 10.1016/B978-0-12-823948-3.00007-5

[45] Ahmad Z, Salman S, Khan SA, Amin A, Rahman ZU, Al-Ghamdi YO, et al. Versatility of hydrogels: From synthetic strategies, classification, and properties to biomedical applications. Gels. 2022;**8**(3):167. DOI: 10.3390/gels8030167

[46] Hoffman AS. Hydrogels for biomedical applications. Advanced Drug Delivery Reviews. 2002;**54**(1):3-12. DOI: 10.1016/s0169-409x(01)00239-3

[47] Revete A, Aparicio A, Cisterna BA, Revete J, Luis L, Ibarra E, et al. Advancements in the use of hydrogels for regenerative medicine: Properties and biomedical applications. International Journal of Biomaterials. 2022;**2022**:3606765. DOI: 10.1155/2022/3606765

[48] Chelu M, Musuc AM, Aricov L, Ozon EA, Iosageanu A, Stefan LM, et al. Antibacterial aloe vera based biocompatible hydrogel for use in dermatological applications. International Journal of Molecular Sciences. 2023;**24**:3893. DOI: 10.3390/ijms24043893

[49] Kopeček J. Hydrogels from soft contact lens and implants to self-assembled nanomaterials. Journal of Polymer Science Part A: Polymer Chemistry. 2009;**47**(22):5929-5946. DOI: 10.1002/pola.23607

[50] Dong YC, Bouché M, Uman S, Burdick JA, Cormode DP. Detecting and monitoring hydrogels with medical imaging. ACS Biomaterials Science & Engineering. 2021;**7**(9):4027-4047. DOI: 10.1021/acsbiomaterials.0c01547

[51] Niazi SK. Non-invasive drug delivery across the blood-brain barrier: A prospective analysis. Pharmaceutics. 2023;**15**(11):2599. DOI: 10.3390/pharmaceutics15112599

[52] Yang F, Shi K, Yp J, et al. Advanced biomaterials for cancer immunotherapy. Acta Pharmacologica Sinica. 2020;**41**:911-927. DOI: 10.1038/s41401-020-0372-z

[53] Chaudhary S, Chakraborty E. Hydrogel based tissue engineering and its future applications in personalized disease modeling and regenerative therapy. Beni-Suef University Journal of Basic and Applied Sciences. 2022;**11**(1):3. DOI: 10.1186/s43088-021-00172-1

[54] Tiwari G, Tiwari R, Sriwastawa B, Bhati L, Pandey S, Pandey P, et al. Drug delivery systems: An updated review. International Journal of Pharmaceutical Investigation. 2012;**2**(1):2-11. DOI: 10.4103/2230-973X.96920

[55] Hu B, Gao J, Lu Y, Wang Y. Applications of degradable hydrogels in novel approaches to disease treatment and new modes of drug delivery. Pharmaceutics. 2023;**15**:2370. DOI: 10.3390/pharmaceutics15102370

[56] Barbero CA. Functional materials made by combining hydrogels (cross-linked polyacrylamides) and conducting polymers (polyanilines)—A critical review. Polymers. 2023;**15**:2240. DOI: 10.3390/polym15102240

[57] Lee S, Tong X, Yang F. Effects of the poly(ethylene glycol) hydrogel crosslinking mechanism on protein release. Biomaterials Science. 2016;**4**(3):405-411. DOI: 10.1039/c5bm00256g

[58] Xu X, Liu Y, Fu W, Yao M, Ding Z, Xuan J, et al. Poly(N-isopropylacrylamide)-based thermoresponsive composite hydrogels for biomedical applications. Polymers. 2020;**12**:580. DOI: 10.3390/polym12030580

[59] Malektaj H, Drozdov AD, deClaville CJ. Mechanical properties of alginate hydrogels cross-linked with multivalent cations. Polymers. 2023;**15**:3012. DOI: 10.3390/polym15143012

[60] Rosiak P, Latanska I, Paul P, Sujka W, Kolesinska B. Modification of alginates to modulate their physic-chemical properties and obtain biomaterials with different functional properties. Molecules. 2021;**26**:7264. DOI: 10.3390/molecules26237264

[61] Gulrez SKH, Al-Assaf S, Phillips GO. Hydrogels: Methods of preparation, characterisation and applications [internet]. In: Progress in Molecular and Environmental Bioengineering - from Analysis and Modeling to Technology Applications. London, UK: InTech; 2011. DOI: 10.5772/24553

[62] Lee JH. Injectable hydrogels delivering therapeutic agents for disease treatment and tissue engineering. Biomaterials Research. 2018;**22**:27. DOI: 10.1186/s40824-018-0138-6

[63] Magana JR, Sproncken CCM, Voets IK. On complex coacervate core micelles: Structure-function perspectives. Polymers. 2020;**12**:1953. DOI: 10.3390/polym12091953

[64] Abuwatfa WH, Awad NS, Pitt WG, Husseini GA. Thermosensitive polymers and thermo-responsive liposomal drug delivery systems. Polymers (Basel). 2022;**14**(5):925. DOI: 10.3390/polym14050925

[65] Walker M. Human skin through the ages. International Journal of Pharmaceutics. 2022;**622**:121850. DOI: 10.1016/j.ijpharm.2022.121850

[66] Abdo JM, Sopko NA, Milner SM. The applied anatomy of human skin: A model for regeneration. Wound Medicine. 2020;**28**:100179. DOI: 10.1016/j.wndm.2020.100179

[67] Kumar A, Sood A, Agrawal G, Thakur S, Thakur VK, Tanaka M, et al. Polysaccharides, proteins, and synthetic polymers based multimodal hydrogels for various biomedical applications: A review. International Journal of Biological Macromolecules. 2023;**247**:125606. DOI: 10.1016/j.ijbiomac.2023.125606

[68] Ijaz F, Tahir HM, Ali S, Ali A, Khan HA, Muzamil A, et al. Biomolecules based hydrogels and their potential biomedical applications: A comprehensive review. International Journal of Biological Macromolecules. 2023;**253**(6):127362. DOI: 10.1016/j.ijbiomac.2023.127362

[69] Sghier K, Mur M, Veiga F, Paiva-Santos AC, Pires PC. Novel therapeutic hybrid systems using hydrogels and nanotechnology: A focus

on nanoemulgels for the treatment of skin diseases. Gels. 2024;**10**(1):45. DOI: 10.3390/gels10010045

[70] Chelu M, Musuc AM. Natural biological macromolecules for designing hydrogels as health care and anti-aging solutions. Engineering Proceedings. 2023;**56**(1):158. DOI: 10.3390/ASEC2023-16519

[71] Bercea M. Bioinspired hydrogels as platforms for life-science applications: Challenges and opportunities. Polymers. 2022;**14**(12):2365. DOI: 10.3390/polym14122365

[72] Tatarusanu S-M, Lupascu F-G, Profire B-S, Szilagyi A, Gardikiotis I, Iacob A-T, et al. Modern approaches in wounds management. Polymers. 2023;**15**(17):3648. DOI: 10.3390/polym15173648

[73] Káčerová S, Muchová M, Doudová H, Münster L, Hanulíková B, Valášková K, et al. Chitosan/dialdehyde cellulose hydrogels with covalently anchored polypyrrole: Novel conductive, antibacterial, antioxidant, immunomodulatory, and anti-inflammatory materials. Carbohydrate Polymers. 2024;**327**:121640. DOI: 10.1016/j.carbpol.2023.121640

[74] Ouyang Y, Su X, Zheng X, Zhang L, Chen Z, Yan Q, et al. Mussel-inspired "all-in-one" sodium alginate/carboxymethyl chitosan hydrogel patch promotes healing of infected wound. International Journal of Biological Macromolecules. 2024;**261**(2):129828. DOI: 10.1016/j.ijbiomac.2024.129828

[75] Chelu M, Popa M, Ozon EA, Pandele Cusu J, Anastasescu M, Surdu VA, et al. High-content Aloe vera based hydrogels: Physicochemical and pharmaceutical properties. Polymers. 2023;**15**(5):1312. DOI: 10.3390/polym15051312

[76] Laurén I, Farzan A, Teotia A, Lindfors NC, Seppälä J. Direct ink writing of biocompatible chitosan/non-isocyanate polyurethane/cellulose nanofiber hydrogels for wound-healing applications. International Journal of Biological Macromolecules. 2024;**259**(2):129321. DOI: 10.1016/j.ijbiomac.2024.129321

[77] Li W, Zheng Y, Pang W, Lai P. Bio-inspired adhesive hydrogel for wound healing. Biomedical Technology. 2023;**1**:65-72. DOI: 10.1016/j.bmt.2022.11.009

[78] Russo E, Villa C. Poloxamer hydrogels for biomedical applications. Pharmaceutics. 2019;**11**(12):67. DOI: 10.3390/pharmaceutics11120671

[79] Salzlechner C, Haghighi T, Huebscher I, Walther AR, Schell S, Gardner A, et al. Adhesive hydrogels for maxillofacial tissue regeneration using minimally invasive procedures. Advanced Healthcare Materials. 2020;**9**(4):1901134. DOI: 10.1002/adhm.201901134

[80] Jumelle C, Yung A, Sani ES, Taketani Y, Gantin F, Bourel L, et al. Development and characterization of a hydrogel-based adhesive patch for sealing open-globe injuries. Acta Biomaterialia. 2022;**137**:53-63. DOI: 10.1016/j.actbio.2021.10.021

[81] Jahan I, Ganesan V, Sahu M, Nandave M, Sen S. Adhesivity-tuned bioactive gelatin/gellan hybrid gels drive efficient wound healing. International Journal of Biological Macromolecules. 2024;**254**:127735. DOI: 10.1016/j.ijbiomac.2023.127735

[82] Glingasorn B, Yongsapanan N, Pangon A, Lin C, Ummartyotin S. Synthesis of bioinspired based hydrogel composite from hyaluronic acid/

polyacrylic acid and lignin as an adhesive for medical technology. Emergent Materials. 2024;**7**:275-284. DOI: 10.1007/s42247-023-00592-y

[83] Ding C, Liu X, Zhang S, Sun S, Yang J, Chai G, et al. Multifunctional hydrogel bioscaffolds based on polysaccharide to promote wound healing: A review. International Journal of Biological Macromolecules. 2024;**259**(Pt 2):129356. DOI: 10.1016/j.ijbiomac.2024.129356

[84] Gonzalez AC, Costa TF, Andrade ZA, Medrado AR. Wound healing - a literature review. Anais Brasileiros de Dermatologia. 2016;**91**(5):614-620. DOI: 10.1590/abd1806-4841.20164741

[85] Diller RB, Tabor AJ. The role of the extracellular matrix (ECM) in wound healing: A review. Biomimetics. 2022;**7**(3):87. DOI: 10.3390/biomimetics7030087

[86] Ozhava D, Bektas C, Lee K, Jackson A, Mao Y. Human mesenchymal stem cells on size-sorted gelatin hydrogel microparticles show enhanced *in vitro* wound healing activities. Gels. 2024;**10**(2):97. DOI: 10.3390/gels10020097

[87] Zarei F, Soleimaninejad M. Role of growth factors and biomaterials in wound healing, artificial cells. Nanomedicine, and Biotechnology. 2018;**46**(1):906-911. DOI: 10.1080/21691401.2018.1439836

[88] Yamakawa S, Hayashida K. Advances in surgical applications of growth factors for wound healing. Burn Trauma. 2019;**7**:10. DOI: 10.1186/s41038-019-0148-1

[89] Park JW, Hwang SR, Yoon IS. Advanced growth factor delivery systems in wound management and skin regeneration. Molecules. 2017;**22**(8):1259. DOI: 10.3390/molecules22081259

[90] Cui G, Guo X, Su P, Zhang T, Guan J, Wang C. Mussel-inspired nanoparticle composite hydrogels for hemostasis and wound healing. Frontiers in Chemistry. 2023;**11**:1154788. DOI: 10.3389/fchem.2023.1154788

[91] Abdel-Gawad R, Osman R, Awad GAS, Mortada N. Wound healing potential of silver nanoparticles embedded in optimized bio-inspired hybridized chitosan soft and dry hydrogel. Carbohydrate Polymers. 2024;**324**:121526. DOI: 10.1016/j.carbpol.2023.121526

[92] Kuhn T, Aljohmani A, Frank N, Zielke L, Mehanny M, Laschke MW, et al. A cell-free, biomimetic hydrogel based on probiotic membrane vesicles ameliorates wound healing. Journal of Controlled Release. 2024;**365**:969-980. DOI: 10.1016/j.jconrel.2023.12.011

[93] Nejati O, Tışlı B, Yaşayan G, Zaman BT, Torkay G, Dönmez M, et al. Microwave-assisted hydrothermal green synthesis of selenium nanoparticles incorporated with hyaluronic acid methacrylate/gelatin methacrylate hydrogels for wound healing applications. Polymer Engineering and Science. 2024;**64**:316-327. DOI: 10.1002/pen.26549

[94] Lee M, Kim YS, Park J, Choe G, Lee S, Kang BG, et al. A paintable and adhesive hydrogel cardiac patch with sustained release of ANGPTL4 for infarcted heart repair. Bioactive Materials. 2023;**31**:395-407. DOI: 10.1016/j.bioactmat.2023.08.020

[95] Korting HC, Schöllmann C, White RJ. Management of minor acute cutaneous

wounds: Importance of wound healing in a moist environment. Journal of the European Academy of Dermatology and Venereology. 2011;**25**(2):130-137. DOI: 10.1111/j.1468-3083.2010.03775.x

[96] Koehler J, Brandl FP, Goepferich AM. Hydrogel wound dressings for bioactive treatment of acute and chronic wounds. European Polymer Journal. 2018;**100**:1-11. DOI: 10.1016/j.eurpolymj.2017.12.046

[97] Reyes-Martínez JE, Ruiz-Pacheco JA, Flores-Valdéz MA, Elsawy MA, Vallejo-Cardona AA, Castillo-Díaz LA. Advanced hydrogels for treatment of diabetes. Journal of Tissue Engineering and Regenerative Medicine. 2019;**13**(8):1375-1393. DOI: 10.1002/term.2880

[98] Shang S, Zhuang K, Chen J, Zhang M, Jiang S, Li WA. A bioactive composite hydrogel dressing that promotes healing of both acute and chronic diabetic skin wounds. Bioactive Materials. 2024;**34**:298-310. DOI: 10.1016/j.bioactmat.2023.12.026

[99] Çelik E, Akelma H. Hydrogel burn dressing effectiveness in burn pain. Burns. 2024;**50**(1):190-196. DOI: 10.1016/j.burns.2023.08.012

[100] Surowiecka A, Strużyna J, Winiarska A, Korzeniowski T. Hydrogels in burn wound management—A review. Gels. 2022;**8**(2):122. DOI: 10.3390/gels8020122

[101] Stoica AE, Chircov C, Grumezescu AM. Hydrogel dressings for the treatment of burn wounds: An up-to-date overview. Materials. 2020;**13**(12):2853. DOI: 10.3390/ma13122853

[102] Li J, Li Y, Guo C, Xiaochen W. Development of quercetin loaded silk fibroin/soybean protein isolate hydrogels for burn wound healing. Chemical Engineering Journal. 2024;**481**:148458. DOI: 10.1016/j.cej.2023.148458

[103] Kumar A, Sood A, Singhmar R, Mishra YK, Thakur VK, Han SS. Manufacturing functional hydrogels for inducing angiogenic–osteogenic coupled progressions in hard tissue repairs: Prospects and challenges. Biomaterials Science. 2022;**10**(19):5472-5497. DOI: 10.1039/D2BM00894G

[104] Salehi Abar E, Vandghanooni S, Torab A, Jaymand M, Eskandani M. A comprehensive review on nanocomposite biomaterials based on gelatin for bone tissue engineering. International Journal of Biological Macromolecules. 2024;**254**(Pt 1):127556. DOI: 10.1016/j.ijbiomac.2023.127556

[105] Lee JS, Kim H-S, Nah H, Moon H-J, Ko NR, Lee SJ, et al. Bioinspired semi-flexible hydrogel with anti-inflammatory potential for natural tissue-mimicking bone regeneration. Composites Part B: Engineering. 2024;**273**:111223. DOI: 10.1016/j.compositesb.2024.111223

[106] Haroon B, Sohail M, Minhas MU, Mahmood A, Hussain Z, Ahmed Shah S, et al. Nano-residronate loaded κ-carrageenan-based injectable hydrogels for bone tissue regeneration. International Journal of Biological Macromolecules. 2023;**251**:126380. DOI: 10.1016/j.ijbiomac.2023.126380

[107] Li Z, Guan J. Hydrogels for cardiac tissue engineering. Polymers. 2011;**3**(2):740-761. DOI: 10.3390/polym3020740

[108] Camci-Unal G, Annabi N, Dokmeci M, et al. Hydrogels for cardiac tissue engineering. NPG Asia Materials. 2014;**6**:e99. DOI: 10.1038/am.2014.19

[109] El-Husseiny HM, Mady EA, El-Dakroury WA, Doghish AS, Tanaka R.

Stimuli-responsive hydrogels: Smart state of-the-art platforms for cardiac tissue engineering. Frontiers in Bioengineering and Biotechnology. 2023;**11**:1174075. DOI: 10.3389/fbioe.2023.1174075

[110] Ye S, Boeter JWB, Penning LC, Spee B, Schneeberger K. Hydrogels for liver tissue engineering. Bioengineering. 2019;**6**(3):59. DOI: 10.3390/bioengineering6030059

[111] Nicodemus GD, Bryant SJ. Cell encapsulation in biodegradable hydrogels for tissue engineering applications. Tissue Engineering. Part B, Reviews. 2008;**14**(2):149-165. DOI: 10.1089/ten.teb.2007.0332

[112] Wu KY, Akbar D, Giunta M, Kalevar A, Tran SD. Hydrogels in ophthalmology: Novel strategies for overcoming therapeutic challenges. Materials. 2024;**17**(1):86. DOI: 10.3390/ma17010086

[113] Aragón-Navas L-CJ, Johnson M, Sigen A, Vicario-de-la-Torre M, Andrés-Guerrero V, Tai H, et al. Smart biodegradable hydrogels: Drug-delivery platforms for treatment of chronic ophthalmic diseases affecting the back of the eye. International Journal of Pharmaceutics. 2024;**649**:123653. DOI: 10.1016/j.ijpharm.2023.123653

[114] Ribeiro AM, Figueiras A, Veiga F. Improvements in topical ocular drug delivery systems: Hydrogels and contact lenses. Journal of Pharmacy & Pharmaceutical Sciences. 2015;**18**(5):683-695. DOI: 10.18433/j3h60p

[115] Arora S, Das G, Alqarni M, Grover V, Manzoor Baba S, Saluja P, et al. Role of chitosan hydrogels in clinical dentistry. Gels. 2023;**9**(9):698. DOI: 10.3390/gels9090698

[116] Chen A, Deng S, Lai J, Li J, Chen W, Varma SN, et al. Hydrogels for oral tissue engineering: Challenges and opportunities. Molecules. 2023;**28**(9):3946. DOI: 10.3390/molecules28093946

[117] Muresan SMC, Dreanca A, Repciuc C, Dejescu C, Rotar O, Pop RA, et al. Dental hydrogels with essential oils with potential activity in periodontitis. Applied Sciences. 2023;**13**(3):1787. DOI: 10.3390/app13031787

[118] Flores-Espinoza AI, Garcia-Contreras R, Guzman-Rocha DA, Aranda-Herrera B, Chavez-Granados PA, Jurado CA, et al. Gelatin–chitosan hydrogel biological, antimicrobial and mechanical properties for dental applications. Biomimetics. 2023;**8**(8):575. DOI: 10.3390/biomimetics8080575

[119] Li X, Xu X, Xu M, Geng Z, Ji P, Liu Y. Hydrogel systems for targeted cancer therapy. Frontiers in Bioengineering and Biotechnology. 2023;**16**(11):1140436. DOI: 10.3389/fbioe.2023.1140436

[120] Cha GD, Lee WH, Sunwoo SH, Kang D, Kang T, Cho KW, et al. Multifunctional injectable hydrogel for *in vivo* diagnostic and therapeutic applications. ACS Nano. 2022;**16**(1):554-567. DOI: 10.1021/acsnano.1c07649

Stimuli-responsive hydrogels: Smart state of-the-art platforms for cardiac tissue engineering. Frontiers in Bioengineering and Biotechnology. 2023;**11**:1174075. DOI: 10.3389/fbioe.2023.1174075

[110] Ye S, Boeter JWB, Penning LC, Spee B, Schneeberger K. Hydrogels for liver tissue engineering. Bioengineering. 2019;**6**(3):59. DOI: 10.3390/bioengineering6030059

[111] Nicodemus GD, Bryant SJ. Cell encapsulation in biodegradable hydrogels for tissue engineering applications. Tissue Engineering. Part B, Reviews. 2008;**14**(2):149-165. DOI: 10.1089/ten.teb.2007.0332

[112] Wu KY, Akbar D, Giunta M, Kalevar A, Tran SD. Hydrogels in ophthalmology: Novel strategies for overcoming therapeutic challenges. Materials. 2024;**17**(1):86. DOI: 10.3390/ma17010086

[113] Aragón-Navas L-CJ, Johnson M, Sigen A, Vicario-de-la-Torre M, Andrés-Guerrero V, Tai H, et al. Smart biodegradable hydrogels: Drug-delivery platforms for treatment of chronic ophthalmic diseases affecting the back of the eye. International Journal of Pharmaceutics. 2024;**649**:123653. DOI: 10.1016/j.ijpharm.2023.123653

[114] Ribeiro AM, Figueiras A, Veiga F. Improvements in topical ocular drug delivery systems: Hydrogels and contact lenses. Journal of Pharmacy & Pharmaceutical Sciences. 2015;**18**(5):683-695. DOI: 10.18433/j3h60p

[115] Arora S, Das G, Alqarni M, Grover V, Manzoor Baba S, Saluja P, et al. Role of chitosan hydrogels in clinical dentistry. Gels. 2023;**9**(9):698. DOI: 10.3390/gels9090698

[116] Chen A, Deng S, Lai J, Li J, Chen W, Varma SN, et al. Hydrogels for oral tissue engineering: Challenges and opportunities. Molecules. 2023;**28**(9):3946. DOI: 10.3390/molecules28093946

[117] Muresan SMC, Dreanca A, Repciuc C, Dejescu C, Rotar O, Pop RA, et al. Dental hydrogels with essential oils with potential activity in periodontitis. Applied Sciences. 2023;**13**(3):1787. DOI: 10.3390/app13031787

[118] Flores-Espinoza AI, Garcia-Contreras R, Guzman-Rocha DA, Aranda-Herrera B, Chavez-Granados PA, Jurado CA, et al. Gelatin–chitosan hydrogel biological, antimicrobial and mechanical properties for dental applications. Biomimetics. 2023;**8**(8):575. DOI: 10.3390/biomimetics8080575

[119] Li X, Xu X, Xu M, Geng Z, Ji P, Liu Y. Hydrogel systems for targeted cancer therapy. Frontiers in Bioengineering and Biotechnology. 2023;**16**(11):1140436. DOI: 10.3389/fbioe.2023.1140436

[120] Cha GD, Lee WH, Sunwoo SH, Kang D, Kang T, Cho KW, et al. Multifunctional injectable hydrogel for *in vivo* diagnostic and therapeutic applications. ACS Nano. 2022;**16**(1):554-567. DOI: 10.1021/acsnano.1c07649

Stimuli-responsive hydrogels: Smart state of-the-art platforms for cardiac tissue engineering. Frontiers in Bioengineering and Biotechnology. 2023;**11**:1174075. DOI: 10.3389/fbioe.2023.1174075

[110] Ye S, Boeter JWB, Penning LC, Spee B, Schneeberger K. Hydrogels for liver tissue engineering. Bioengineering. 2019;**6**(3):59. DOI: 10.3390/bioengineering6030059

[111] Nicodemus GD, Bryant SJ. Cell encapsulation in biodegradable hydrogels for tissue engineering applications. Tissue Engineering. Part B, Reviews. 2008;**14**(2):149-165. DOI: 10.1089/ten.teb.2007.0332

[112] Wu KY, Akbar D, Giunta M, Kalevar A, Tran SD. Hydrogels in ophthalmology: Novel strategies for overcoming therapeutic challenges. Materials. 2024;**17**(1):86. DOI: 10.3390/ma17010086

[113] Aragón-Navas L-CJ, Johnson M, Sigen A, Vicario-de-la-Torre M, Andrés-Guerrero V, Tai H, et al. Smart biodegradable hydrogels: Drug-delivery platforms for treatment of chronic ophthalmic diseases affecting the back of the eye. International Journal of Pharmaceutics. 2024;**649**:123653. DOI: 10.1016/j.ijpharm.2023.123653

[114] Ribeiro AM, Figueiras A, Veiga F. Improvements in topical ocular drug delivery systems: Hydrogels and contact lenses. Journal of Pharmacy & Pharmaceutical Sciences. 2015;**18**(5):683-695. DOI: 10.18433/j3h60p

[115] Arora S, Das G, Alqarni M, Grover V, Manzoor Baba S, Saluja P, et al. Role of chitosan hydrogels in clinical dentistry. Gels. 2023;**9**(9):698. DOI: 10.3390/gels9090698

[116] Chen A, Deng S, Lai J, Li J, Chen W, Varma SN, et al. Hydrogels for oral tissue engineering: Challenges and opportunities. Molecules. 2023;**28**(9):3946. DOI: 10.3390/molecules28093946

[117] Muresan SMC, Dreanca A, Repciuc C, Dejescu C, Rotar O, Pop RA, et al. Dental hydrogels with essential oils with potential activity in periodontitis. Applied Sciences. 2023;**13**(3):1787. DOI: 10.3390/app13031787

[118] Flores-Espinoza AI, Garcia-Contreras R, Guzman-Rocha DA, Aranda-Herrera B, Chavez-Granados PA, Jurado CA, et al. Gelatin–chitosan hydrogel biological, antimicrobial and mechanical properties for dental applications. Biomimetics. 2023;**8**(8):575. DOI: 10.3390/biomimetics8080575

[119] Li X, Xu X, Xu M, Geng Z, Ji P, Liu Y. Hydrogel systems for targeted cancer therapy. Frontiers in Bioengineering and Biotechnology. 2023;**16**(11):1140436. DOI: 10.3389/fbioe.2023.1140436

[120] Cha GD, Lee WH, Sunwoo SH, Kang D, Kang T, Cho KW, et al. Multifunctional injectable hydrogel for *in vivo* diagnostic and therapeutic applications. ACS Nano. 2022;**16**(1):554-567. DOI: 10.1021/acsnano.1c07649

Chapter 5

Developments and Challenges of Hydrogel Coatings for Long-Term Marine Antifouling Applications

Mengyan Nie and Koulin Chen

Abstract

As a result of the accumulation of marine organisms on submerged surfaces, marine fouling can have significant economic and environmental impacts. For example, marine fouling can increase drag and reduce the hydrodynamic efficiency of a vessel, leading to increased fuel consumption and operational costs as well as higher greenhouse gas emissions. The marine organisms attached to submerged surfaces can also induce corrosion and cause the marine structural integrity of the affected surfaces compromised, leading to increased maintenance and repair costs. Additionally, marine fouling can also pose biosecurity risks by spreading invasive species to new regions and disrupting local ecosystems. Great efforts have been made to develop effective and environmentally friendly antifouling technologies to mitigate these impacts. Hydrogel antifouling coatings have been proven effective and environmentally friendly, making them promising for practical marine applications. Here, brief overviews of antifouling mechanisms and types of hydrogel coatings are presented first. The latest developments in hydrogel antifouling coatings are categorized based on design strategies, and the limitations of these coatings are also critically appreciated with regard to their potential for practical marine applications. Finally, insightful perspectives on hydrogel coating are summarized for their use in practical marine applications.

Keywords: hydrogel, hydration layer, biofouling, coating, marine fouling, microbiologically induced corrosion (MIC), antibacterial, coating toughness, self-generating, self-healing, antifouling agent

1. Introduction

Material surfaces constantly face the threat of foreign substance adherence. When materials become covered in contaminants, their inherent interfacial properties are compromised, which would cause severe impact on their performance and even failure of engineering components. Microbial biofouling poses a significant issue in relation to biomedical devices, implants, marine structures, and food packaging materials. The formation of microbial biofilms on biomedical devices and implants can lead to severe infections at surgical sites and even medical devices premature failure. As such, medical devices in contact with human blood must prevent protein adsorption and blood cell adhesion to avert biofilm formation and biofouling. In aquatic environments, microbes, macroalgae, and shellfish readily attach to alloys and metals, causing microbiologically induced corrosion (MIC), reducing fuel efficiency and operational performance in naval vessels, and contributing to transregional biological invasions [1].

To tackle these challenges, various antimicrobial strategies have been developed to effectively prevent bacterial attachment, broadly categorized into two main approaches: 'contact killing' which actively kills adhering bacteria and 'antifouling surfaces' that resist bacterial attachment. The approach of contact-killing often leads the killed bacteria and debris to accumulate on the surface, thereby compromising the long-term antibacterial performance of the system or the effective surface regeneration strategies needed. The antifouling surface approach mainly relies on surface modification strategies to form antifouling coatings or to produce specific surface morphologies [2]. Among current antifouling coatings, hydrogels comprising 3D networks of soft and wet materials are more promising due to their high antifouling performance and environmental friendliness. In this chapter, the working mechanisms of hydrogel antifouling coatings are briefly overviewed first, and then different types of antifouling hydrogels are critically appreciated with their limitations identified and mitigation strategies proposed. In the end, the perspectives on hydrogel antifouling coating systems are presented to address the remaining challenges for practical marine applications.

2. Antifouling mechanism of hydrogels

Due to a variety of amino-acid residues and conformation flexibility, proteins could adsorb onto nearly any material surface via noncovalent interactions, such as hydrogen bonds, electrostatic and ionic interactions, and hydrophobic interactions. Cells do not directly attach to any substances, rather they interact with the materials via the adsorbed proteins. The conformation of the adsorbed proteins dictates how the cells respond (e.g., adhesion, proliferation, differentiation, etc.). In order to prevent protein adsorption and bacterial attachment, two strategies are commonly used to fabricate antifouling coatings—to make the surface either strongly hydrated or hydrophobic. For example, the perfluorinated surfaces are usually super-hydrophobic and oleophobic, exhibiting good antifouling performance because of low-affinity interactions between most molecules and the superhydrophobic surface. By contrast, the hydration layer would form a physical and energy barrier to prevent molecule adsorption. A wide range of molecule systems with hydrophilic or ionic groups that can form strongly hydrated layers have been employed to fabricate antifouling surfaces, such as poly(ethylene glycol) (PEG) and its derivatives, zwitterionic materials, peptides, polysaccharides, and other polymers [3].

Hydrogels are composed of hydrophilic polymer networks. Their super-hydrophilic nature allows them to absorb significant amounts of water molecules into their 3D networks, forming a highly hydrated layer on their surface that prevents the adhesion of proteins or microbes. For an organism to attach, it must penetrate this hydration layer, requiring additional energy and thus reducing the likelihood of attachment [4, 5]. Moreover, the swollen state of hydrogels imparts soft and highly elastic properties, deterring most marine organisms that prefer to adhere to rigid surfaces. Consequently, hydrogels exhibit excellent resistance against proteins, marine bacteria, green algae spores, and barnacle larvae. They are also nontoxic, highly elastic, and inert to bio-macromolecular adhesion.

Hydrogels, including polyhydrophilic and zwitterionic polymers, exhibit different mechanisms for forming their hydration layers. Hydrophilic materials leverage hydrogen bonding while zwitterionic materials utilize stronger ionic solvation. Resistance to protein adsorption correlates positively with the strength of surface hydration, which is primarily determined by the material's physicochemical properties (i.e., molecular weight and surface chemistry) and surface accumulation (i.e., film thickness, packing density, and chain conformation). In addition to surface hydration, chain flexibility plays a crucial role in protein resistance, especially for long-chain polymers. For instance, UV-initiated thiol-ene cross-linking chemistry was utilized to prepare PEG hydrogel coatings with different structural compositions by varying PEG length, vinylic end group, and thiol cross-linker [6]. The antifouling efficiency of long-chain PEG outperformed that of oligo(ethylene glycol) coatings. As proteins approach the surface, the compression of polymer chains due to entropy reduction leads to steric repulsion, resisting protein adsorption, and marine bacteria settlement. While most water-soluble polymers can reduce protein adsorption to some extent, optimal antifouling capabilities are achieved only under the combined effects of surface hydration and steric repulsion. It is speculated that the hydration layer plays the main role in the resistance against protein adsorption if chain flexibility is limited, whereas both chain flexibility and the hydration layer are important if chain flexibility is significant [4].

3. Types of hydrogels used for antifouling coatings

3.1 Polyhydrophilic polymer hydrogels

Several low- or nonfouling polyhydrophilic materials, including poly(ethylene glycol) (PEG, sometimes termed as polyethylene oxide, PEO), polysaccharides, and polyamides, share common structural and chemical properties—hydrophilicity, electrical charge neutrality, and the presence of hydrogen-bond acceptors/donors [4].

PEG and its derivatives are the most widely used antifouling materials due to the advantages of high solubility in water, superior biocompatibility, and nontoxicity. Since the 1970s, they have been utilized to modify material surfaces for antifouling coatings because of their strong antifouling tendencies toward cells and proteins. Due to their bulky volume, high hydrated chain mobility, and steric repulsion, PEGylated materials effectively reduce cell adhesion and protein adsorption [7]. The higher the degree of polymerization, the better the antifouling effect is. Ekblad et al. [8] have demonstrated that hydrogel coatings containing polyethylene glycol are highly effective in inhibiting the settlement of a wide range of fouling organisms in marine and freshwater, showing excellent antifouling properties with regard to settlement and removal. However, their widespread use is hindered by poor mechanical performance and brittleness upon dehydration [9, 10]. Furthermore, PEG undergoes oxidative degradation when exposed to oxygen, elevated temperatures, or light [11, 12], and the degradation process is significantly accelerated when ethylene glycol segments come into direct contact with a transitional metal catalyst such as gold, in the presence of oxygen and higher temperatures [13]. The oxidative degradation would make the PEGylated surface unstable and diminish their antifouling properties, and thus limiting their long-term applications. For example, PEG brushes lose their antifouling capabilities when temperatures reach 35°C, limiting their long-term resistance to protein fouling [14].

3.2 Zwitterionic polymer hydrogels

Compared with PEG and its derivatives, zwitterionic polymer materials, such as carboxybetaine (CB), sulfobetaine (SB), and phosphorylcholine (PC) demonstrate greater oxidative resistance and hydrolytic stability, and have been explored extensively as promising alternatives to PEG derivatives in developing antifouling surfaces. Zwitterionic polymers, such as polybetaines and polyampholytes, contain both positive and negative charges equally within their structure. Unlike nonionic hydrophilic polymers (e.g., PEG) that form hydration layers through hydrogen bonding, zwitterionic polymers can form a more stable and thicker hydration layer on the surface via electrostatic interactions, tightly binding water molecules through more effective and stable ionic bonds (as seen in **Figure 1**) [15], which reduce electrostatic interactions with protein molecules and minimize protein adhesion, yielding superior antifouling performance. In addition, the electrostatic interactions between water modules and dipoles present in the zwitterionic polymer chain are stronger and less sensitive to temperature than hydrogen-bonding interactions along EG chains, thus as-induced hydration layer more stable at higher temperatures (e.g. body temperature).

A key factor in the nonfouling properties of zwitterionic polymers is to control the uniform distribution of surface charge and to achieve surface charge neutrality (**Figure 2**), guiding the design of novel, fouling-resistant zwitterionic materials. Balanced charge and minimal dipole moments are crucial for enhancing the hydration capacity of the outermost layer through electrostatic interactions [16–20].

However, zwitterionic polymers swell in aqueous media, leading to poor adhesion to surfaces. Therefore, to create stable zwitterionic surfaces, extensive surface modification and covalent bonding techniques are required. Moreover, external conditions such as pH, ionic strength, and temperature significantly influence the antifouling performance of zwitterionic materials. All these factors may limit their ultimate usability in industrial applications [21].

Based on their negatively charged groups, polybetaines are categorized into sulfobetaine (SB), carboxybetaine (CB), and phosphobetaine (PB). They can be considered biomimetic molecules due to their structural similarity to natural molecules. The structure of SB is analogous to taurine, a sulfur-containing nonprotein amino acid widely present in many animals [22]. CB resembles glycine betaine, a natural compound used in humans for osmoregulation [23]. PB is homologous to the hydrophilic

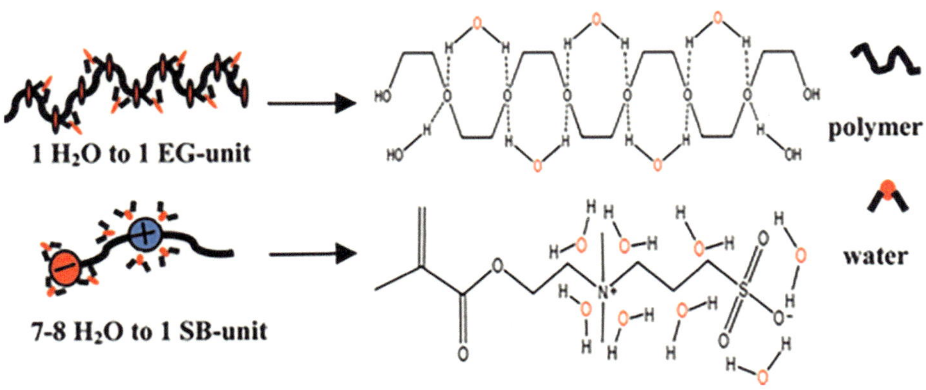

Figure 1.
Comparison of the hydration of poly(sulfobetaine methacrylate) (PSBMA) and poly(ethylene glycol) (PEG) [15].

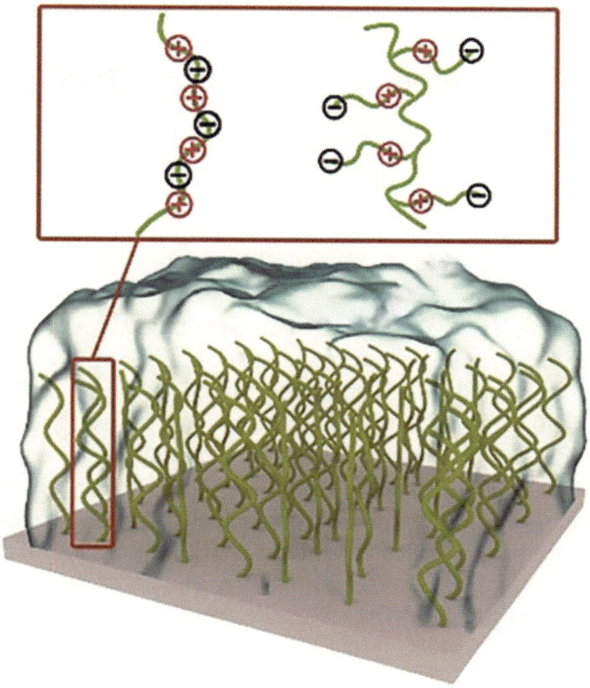

Figure 2.
The uniform distribution of surface charge of polybetaines and polyampholytes [16].

portion of phospholipids, which are major components of cell membranes [24]. However, due to high production costs, PB-based polymers have limited their applications for antifouling coatings. In contrast, SB and CB monomers are cost-effective in production. Especially, SB can be easily prepared, and the antifouling performance of polysulfobetaine is unaffected by the medium solution pH level. CB monomers are biocompatible and can be further modified to enhance their antifouling performance. Therefore, current antifouling coatings primarily utilize SB and CB [25]. Karthäuser et al. [26] synthesized a set of poly(sulfobetaine methacrylate)s (PSBMAs) to study the relationship between their structure and antifouling performance. The spacer groups, either separating the zwitterionic units from the polymer backbone or the cationic from the anionic charges, were systematically varied to explore its effect on the resistance against nonspecific protein adsorption and the accumulation of single species of marine biofouling organisms. It was revealed that the antifouling performance of PSBMAs coating could be optimized via the design of the betaine-to-backbone spacers. The shorter the ethylene spacer, the higher protein resistance is achieved. Moreover, a shorter spacer between the oppositely charged ionic groups of the zwitterionic moiety effectively favors the removal of tested fouling organisms.

In addition to polybetaines, polyampholytes represent another class of materials that are structurally similar to polybetaine polymers. These polymers achieve uniform charge distribution and charge neutrality primarily through the equimolar and homogenous pre-polymerization mixture of two monomers bearing opposite charges. Ampholytic components are notably amenable to functionalization, offering broader chemical diversity and greater flexibility in molecular design. Yeon et al. [27] synthesized an ampholytic dopamine derivative, ZW-DOPA, which contains both catechol and amine

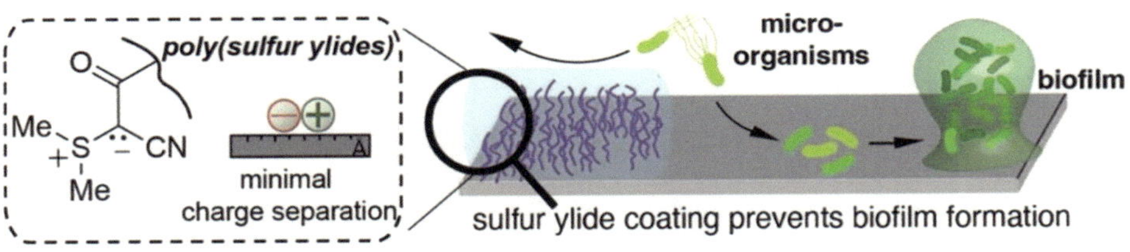

Figure 3.
Illustration of poly(sulfur ylides) coatings for preventing adsorption of biomolecules and microorganisms [29].

functionalities. Coatings based on ZW-DOPA demonstrated enhanced resistance to adhesion by marine diatoms. Furthermore, Liu et al. [28] developed a new subsurface-initiated atom transfer radical (SIATR) polymerization technique with sulfobetaine methacrylate and 3-sulfopropyl methacrylate potassium salt as monomers to fabricate ampholytic copolymer brushes, which demonstrated significantly enhanced stability of the ampholytic polymers and promising candidate for long-term antifouling coatings in marine environment. Berking et al. [29] explored a new class of zwitterionic polymers, specifically poly(sulfur ylides), which effectively inhibit the adhesion of biomolecules and pathogenic bacteria. It is also noteworthy that minimal charge separation exists in these new zwitterionic polymers (as seen in **Figure 3**), which agrees well with the favorable shorter spacer effect observed by Karthäuser et al. in [26], providing insightful guidance for the development of better zwitterionic polymers antifouling coatings.

Zhang et al. [30] also developed zwitterionic hyaluronic acid-based hydrogels via UV light-free thiol-ene click chemistry reaction, which exhibited excellent antifouling properties and good cytocompatibility and were well suitable for 3D cell encapsulation.

4. Strategies for improving the antifouling performance of hydrogel for marine applications

While the hydrophilic surfaces of hydrogels help prevent fouling by reducing the adhesion of proteins and microorganisms, their antifouling applications are still limited by unsatisfactory antimicrobial properties. The hydration layer of zwitterionic hydrogel coatings is sensitive to pH, ionic strength, and temperature of the exposed environment, which could significantly influence antifouling behaviors of zwitterionic coatings and affect protection naval vessels from biofouling when operating in different locations or seasons. Furthermore, some hydration layer-based coatings may detach or de-graft from the surfaces due to the hydrolysis of siloxane, amide, or ester bonds or cleavage of Au–S or oxidative degradation of PEG in the hydration layer. Various approaches have been developed to improve the mechanical, durability, and antifouling performances of hydrogel coatings for long-term applications in marine environments.

4.1 Enhancing adhesion to substrates via surface grafting or synergistic multimodal hydrogen bonding

It remains challenging to stably and conveniently deposit hydrophilic polymers densely and homogeneously onto various surfaces. Especially in marine antifouling, large-scale applications of hydrophilic polymer coatings can lead to problems, such as low adhesive strength, uneven surface coverage, and short lifespan. Effectively anchoring PEG to material surfaces poses a significant challenge. Various methods, including physical adsorption [31], graft polymerization [32, 33], covalent coupling [34, 35], and plasma treatment, have been employed. For example, Kim et al. [36] synthesized a series of stable PEG-based peptides (PEGtides) hydrophilic coatings via anionic ring-opening polymerization of functional epoxy monomers like catechol and lysine, leveraging catechol–amine synergy and diverse hydrogen bonding for strong adhesion (as seen in **Figure 4(a)**). Wang et al. [37] achieved the strong adhesion of catechol-modified four-arm polyethylene glycol hydrophilic antifouling coating onto various types of materials surfaces by grafting dopamine onto the ends of four-arm PEG (4A-PEG-COOH)

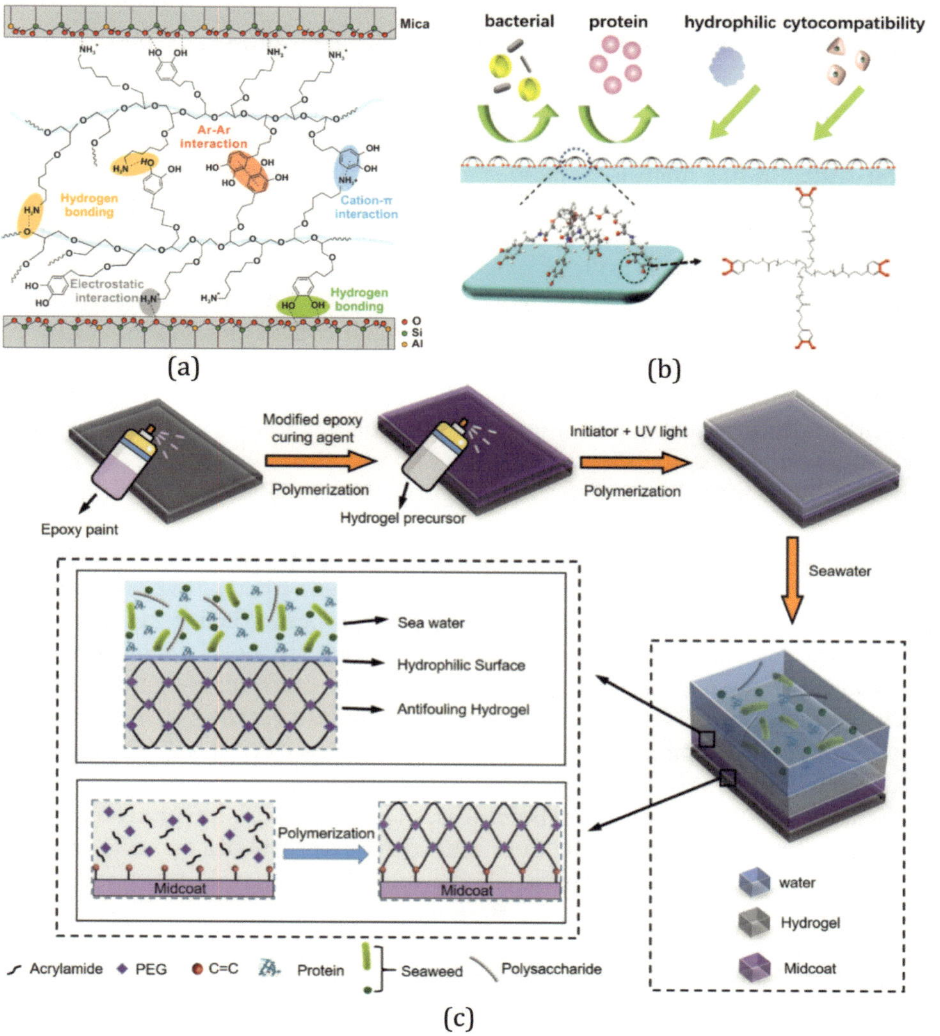

Figure 4.
Schematic diagrams of (a) the overall adhesion and cohesion mechanism of the PEGtides developed in [36], (b) 4A-PEG-DA coating in [37], (c) covalently bonded interlayer introduced between the antifouling hydrogel and substrate surfaces [38]. In (c), a layer of mid-coat is used to mediate the binding of the hydrogel layer to various surfaces. Covalent anchoring groups are used to establish strong adhesion of the hydrogel layer to the mid-coat.

(as seen in **Figure 4(b)**) through amidation reaction. Zhu et al. [39] selected an ethyl α-cyanoacrylate (ECA) adhesive that maintains a robust bond between the polyvinyl alcohol (PVA)-glycerin gel and the substrate through several dehydration/rehydration cycles, thus preserving its antifouling efficacy with excellent thermoplastic and mechanical properties. PVA-glycerol hydrogels significantly inhibit the settlement of the barnacle Balanus albicostatus. Yang et al. [38] introduced an epoxy resin interlayer (mid-coat as called in **Figure 4(c)**) in hydrogel coatings to facilitate strong hydrogel attachment to various surfaces. The strong adhesion of the hydrogel layer to the mid-coat was achieved via the formation of strong covalent bonds between them. Despite covalent bonding, the hydrogel layer provided a hydrophilic surface, thus maintaining outstanding antifouling performance.

4.2 Introducing nanomaterials for nanocomposite antifouling coatings

Some research groups have investigated the antifouling performance of organic/inorganic hybrid composite materials by mixing inorganic fillers at the nano- or micron-scale. Adding micron- or nano-sized fillers can enhance the toughness of the coating, and formation of nanocomposites can significantly improve mechanical properties of the hydrogel coatings. Liu et al. [40] first functionalized multiwalled carbon nanotubes (MWCNTs) with poly(sulfone) (PSf) and sulfobetaine methacrylate (SBMA) to produce amphiphilic and protein-resistant nanocomposite films and membranes. Wang et al. [41] later prepared PSf nano-hybrid membranes enhanced with functionalized MWCNTs with PEG as the dispersion phase, and as-prepared hybrid membranes showed better hydrophilicity and excellent antifouling properties while maintaining excellent tensile strength. Carbon nanotubes, with their excellent mechanical properties, high surface area, and good hydrophilicity, have become an ideal component for enhancing mechanical properties and increasing the porosity of composite membranes.

Ashraf [42] developed a strong and uniform nano-CuO incorporated PEG hydrogel coating over polyaniline-modified polyethylene aquaculture cage nets and observed a significant reduction in fouling over three-month field tests. Tian et al. [43] prepared a series of hybrid antifouling coatings consisting of silicone elastomer and nanocomposite hydrogel embedded with silver nanoparticles (AgNPs). The field tests in the South China Sea demonstrated that the hybrid coatings with nanocomposite hydrogel exhibited good antifouling performance regarding antialgae properties (as shown in **Figure 5**).

Zhang et al. [44] developed a cellulose-based hydrogel with integrated copper oxide nanoparticles, where Cu_2O@CuO nanospheres were in-situ deposited onto the cellulose hydrogel framework. This method ensures better distribution and strong binding between the nanospheres and the hydrogel. The cellulose hydrogel exhibited not only antifouling properties due to the hydration layer on its surface but also enhanced mechanical strength, making it more durable. Li et al. [45] developed nanocomposite hydrogel coatings containing 3D porous Cu_2O nanoparticles, redox gel, and reduced graphene oxide, and observed excellent, stable, and long-lasting bactericidal performance with regard to mussel and barnacle adhesion. Instead of Cu_2O nanoparticles, Xiong et al. [46] dispersed CeO_2 nanorods into a polyzwitterionic hydrogel PVA-P(SBMA-AM) via a simple one-pot method and achieved outstanding antifouling performance, which can sustain for over 6 months in a real marine environment. The highly efficient antifouling is due to a dense hydrated layer formed on the surface

DOI: http://dx.doi.org/10.5772/intechopen.1005794

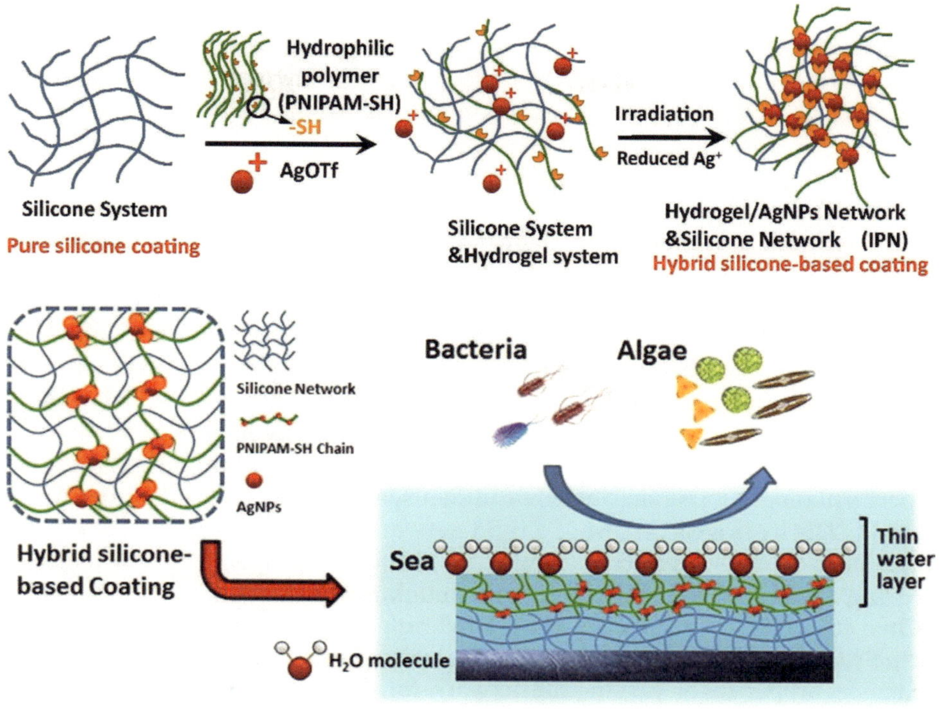

Figure 5.
Schematic illustration of the formation of the pure silicone film and the hybrid coatings (upper) and antifouling performance (lower) [43].

by the super-hydrophilic polyzwitterionic hydrogel and its subsequently excellent broad-spectrum antiadhesion behavior. CeO_2 nanorods dispersed in the hydrogel also have highly efficient bacterial-killing performance. Synergistic work between the bacterial killing of CeO_2 nanorods and antiadhesion of the polyzwitterionic hydrogel leads to its significantly improved marine-antifouling performance. Very recently, Hu et al. [47] employed $Al(OH)_3$ nanoparticle as a physical cross-linker to construct a soft antifouling polyacrylic acid (PAA)/PSBMA hydrogel coating with high toughness and low swelling through dynamic coordination bonding and plentiful hydrogen bonds.

Wei et al. [48] synthesized spindle-shaped calcium carbonate-chitosan/poly (vinyl alcohol) (SCC/PVA) hydrogels, inspired by the characteristics of good mechanical strength and self-cleaning capability of seashells. These hydrogels, containing 1–2-μm-sized spindle $CaCO_3$ particles which are well dispersed within the matrix, demonstrated stable underwater superoleophobicity, remarkable mechanical strength (with compressive strengths up to 25 MPa for SCC/PVA-3), and long-term antibiofouling performance (up to 180 days testing period) against typical fouling organism in marine environment.

Due to their high specific surface area, nanoparticles exhibit a strong tendency to agglomerate, which can negatively impact the mechanical properties of the hydrogels. Although numerous methods have been employed to modify nanoparticles to reduce their aggregation, using excessively modified nanoparticles in hydrogels can still lead to aggregation and decrease the mechanical properties of the hydrogels. Therefore, the optimal dosage of nanoparticles should be thoroughly investigated, and further developments in nanoparticle modifications should be pursued to enhance their dispersion. Furthermore, the aforementioned metallic nano-fillers as fungicides or bactericides, such as AgNPs, Cu_2O, CeO_2, and $Al(OH)_3$, are hazardous to marine ecosystems if overused, and should be replaced by some environmentally compatible materials (e.g., calcium carbonate, chitosan, or their analogs).

4.3 Introducing double-network structure to toughen hydrogels

The mechanical properties of single network (SN) antifouling hydrogels are generally poor. For example, single network poly(sulfobetaine methacrylate) (PSBMA) hydrogel has a low fracture strain of 74.19 ± 4.27% and a weak maximum compressive stress of 0.191 ± 0.019 MPa. The compressive stress and fracture strain of another zwitterionic single network poly(carboxybetaine methacrylate) (PCBMA) hydrogel are 0.531 ± 0.058 MPa and 69.88 ± 1.93%, respectively [49]. According to the literature [43, 46, 50], introducing double network (DN) structures can effectively strengthen hydrogels and improve their long-term antifouling performance in harsh marine environments, as demonstrated in **Figure 6**. Typically, DN hydrogels consist of a rigid network that breaks easily to dissipate energy and a flexible network that enhances the tensile properties of the hydrogels, achieving a balanced state of strength and toughness [51].

Chen et al. [52] designed a new type of hybrid physically-chemically cross-linked agar/polyacrylamide (PAM) double network hydrogel with desirable/balanced mechanical properties by varying the network-forming parameters. Under optimal

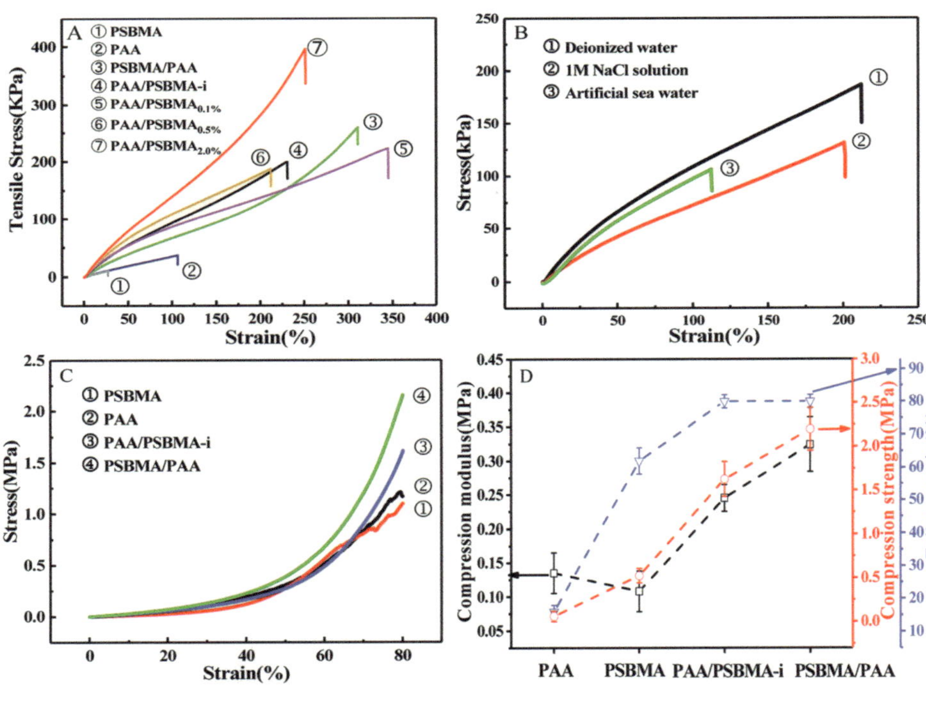

Figure 6.
(a) Tensile stress–strain curves for single-network (PAA and PSBMA) and double-network (PAA/PSBMA and PSBMA/PAA) hydrogels in deionized water. (b) Tensile stress–strain curves for double-network PAA/PSBMA$_{0.5\%}$ hydrogel under deionized water, 1 M NaCl solution and artificial sea water. (c) Compressive stress–strain curves for PAA, PSBMA, PAA/PSBMA-I, and PSBMA/PAA hydrogels. (d) Compressive properties of single- and double-network hydrogels [50].

conditions, agar/PAM DN hydrogels achieved the greatest tensile stress of approximately 3.3 MPa at a failure strain of *ca*. 2400% and the highest tensile strain of 3700% at a failure stress of 2.8 MPa, comparable to the chemically cross-linked DN hydrogels. Agar/PAM hydrogels also exhibited excellent antifouling properties by strongly resisting protein adsorption, cell adhesion, and bacterial attachment, as well as the free shapeable property to form any complex shapes.

Jiang et al. [53] prepared double network hydrogel by copolymerization of different ratios of [2-(methacryloyloxy) ethyl] trimethylammonium (TMA) and 3-sulfopropyl methacrylate (SA), followed by incorporation of a second polyacrylamide (PAM) network. Compared with the single network hydrogels, the elastic modulus, maximum compressive stress, and strain of the DN hydrogels measured by compression tests were enhanced greatly with a second PAM polymer network incorporated. It was also observed from protein adsorption and algae attachment tests that the negatively charged hydrogels showed better antialgae fouling performance than the positively charged and neutral DN hydrogels.

Great efforts have recently been made to develop double-network hydrogel for biocompatible or environment-friendly antibacterial coatings for long-term real applications in biomedical devices and naval vessels. **Table 1** summarizes some innovative approaches reported by far [46, 50, 52–62] for double network hydrogel coatings which show promise for long-term antifouling applications in harsh marine environments, especially for naval assets and reported in the past 5 years. Various interaction modes within and between networks are explored to enhance mechanical properties or to introduce additional functions to prolong the antifouling service life of hydrogel coatings. For example, Zhang et al. [58] employed a simple one-pot method to prepare a hybrid ionic-covalent cross-linked double-network hydrogel. In the DN hydrogel, sodium alginate (SA) was cross-linked by Ca^{2+} ions to serve as the rigid network, while PCBAA was covalently cross-linked to act as the flexible network. This DN hydrogel exhibited excellent mechanical properties, including high elastic modulus (0.28 MPa), high tensile strength (0.69 MPa), and robust self-healing capabilities. More importantly, the hybrid cross-linked double-network hydrogel demonstrated strong resistance to nonspecific protein, cell, bacterial, and algal adhesion, showcasing outstanding antifouling performance. Li et al. [61] developed a DN hydrogel with chemically cross-linked PAM as the first network and physically cross-linked copper alginate as the second network. The introduction of copper ions (Cu^{2+}) into the alginate network endowed the hydrogel with antibacterial properties and provided better mechanical properties through stronger cross-linking between alginate and copper ions compared to traditional calcium ions (Ca^{2+}). The introduction of BTA is believed to endow the hydrogel coating with anticorrosion properties, transforming pitting corrosion into uniform corrosion, and providing unique advantages in tidal zone applications. Still using physically cross-linked sodium alginate as the second network, Xiong et al. [62] developed an ultrahigh strength and outstanding marine antifouling hydrogel coating with chemically cross-linked PVA as the first network while introducing polyhexamethylene biguanide hydrochloride (PHMB) potent fungicide. The as-developed hydrogel demonstrated a high tensile strength of 17.23 MPa and excellent antifouling effect with no attachment observed from any marine organisms on the hydrogel coating after 6 months of immersion in the actual marine environment. The ultrahigh tensile strength of the hydrogels also remained nearly the same after 6 months of immersion in the sea. As such, the as-developed antifouling hydrogels are very promising for real applications in harsh marine environments.

Hydrogels	Interactions in DN	Mechanical properties of DN hydrogel achieved	Antifouling performance	Additional functions	Ref.
Agar/Polyacrylamide (PAM)	Hybrid: H-bond (agar) and covalent cross-linked (PAM)	E = 123 kPa, $\sigma_{f,max}$ = 3.3 MPa, $\varepsilon_{f,max}$ = 3700%	Excellent in different biological media	Free-shapeable property	[52]
Agar/Hydrophobic PAM (HPAM) (via stearyl methacrylate)	Fully physical: H-bond (agar) and hydrophobically associated PAM	E = 106–113 kPa, σ_f = 0.26–0.37 MPa, ε_f = 3390–5260%, W = 7.16–9.96 kJ m^{-3}	Not tested	Self-recovery, self-healing, improved fatigue resistance	[54]
Copolymer [2-(meth-acryloyloxy) ethyl] trimethylammonium (TMA) and 3-sulfopropylmethacrylate (SA)/PAM	Hybrid ionic-covalent cross-linked	E = 2.88 MPa, σ_f = 0.432 MPa at ε_f = 13.3% for DN-7-3 hydrogel (compression test)	Excellent on DN hydrogel with TMA/SA = 1:1	N/A	[53]
Chitosan (CS)/poly-(sulfobetainemethacrylate) (PSBMA)	Hybrid ionic-covalent cross-linked	E = 500 kPa, σ_f = 2.0 MPa, ε_f = ~820%, W = 1.360 MJ m^{-3}	Excellent in vitro and in vivo with 'repel and kill' effect	Fast self-recovery, excellent fatigue resistance	[55]
Polyacrylic acid (PAA)/PSBMA	Hybrid to fully covalently cross-linked	E = ~110 kPa, σ_f = ~0.4 MPa, ε_f = ~350%	Intermediate compared with SN PAA or PSBMA	Low swelling degree, drag-reduction	[50]
CS/(N-(2-hydroxyethyl)acrylamide) (PHEAA)	Hybrid ionic-covalent cross-linked	E = 600 kPa, σ_f = 3.8 MPa, ε_f = ~700%, W = ~ 6.02 MJ m^{-3}	Excellent in vitro with 'repel and kill' effect	Excellent fatigue resistance, fast self-recovery	[56]
Sodium alginate (SA)/PHEAA	Hybrid ionic-covalent cross-linked	E = 310 kPa, σ_f = 1.32 MPa, ε_f = ~700%	Well in vitro	Self-recovery, excellent fatigue resistance	[57]
SA/polycarboxybetaine acrylamide (PCBAA)	Hybrid ionic-covalent cross-linked	E = 280 kPa, σ_f = 0.69 MPa, ε_f = ~400%, W = ~ 500 kJ m^{-3}	Outstanding for marine environments	Good self-recovery, highly biocompatible	[58]
3-(1-(4-Vinylbenzyl)-1H-imidazol-3-ium-3-yl)-propane-1-sulfonate (VBIPS)	Fully physical: Ionic and π-π interactions	E = ~60 kPa, σ_f = ~0.20 MPa, ε_f = ~450%, W = ~ 235 kJ m^{-3}	Excellent in short-term but feasible long-term via salt treatment	Self-healable in acid, pH-sensitive strength, antifreezing	[59]
Polyvinyl alcohol (PVA)/N-(4-hydroxy-3-methoxybenzyl) acrylamide (HMBA)	Hybrid H-bond and covalent cross-linked	σ_f = 3.21 MPa at ε_f = 80% (compression test)	Excellent, long-term marine antifouling	Green antifouling	[60]

DOI: http://dx.doi.org/10.5772/intechopen.1005794

Hydrogels	Interactions in DN	Mechanical properties of DN hydrogel achieved	Antifouling performance	Additional functions	Ref.
Poly(sulfonate betaine-acrylamide) [P(SBMA-AM)]/CeO_2 nanorod-PVA (CeO_2 – PVA)	Hybrid covalent cross-linked P(SBMA-AM) and H-bond cross-linked CeO_2 – PVA	σ_f = 2.44 MPa, ε_f = ~450%, W = 492 kJ m^- σ_f = 27.87 MPa (compression stress)	Outstanding antifouling in real marine environment over 6 months	High-strength, highly effective long-term marine antifouling	[46]
PAM/SA-Benzotriazole (BTA)	Hybrid ionic-covalent cross-linked	σ_f = 200 kPa, W = 128 J m^{-2}	Excellent under marine tidal environment	*Anticorrosion*, long-term antifouling	[61]
Polyvinyl alcohol (PVC)/ SA-polyhexamethylenebiguanide hydrochloride (PHMB)	Hybrid ionic-covalent cross-linked	σ_f = 17.23 MPa, ε_f = 388%,	Excellent over 6 months in marine environment	Ultrahigh strength, outstanding antifouling	[62]

Notes: Elastic modulus (E), Fracture stress (σ_f), dissipated energy (W).

Table 1.
Summary of innovative approaches reported for double network (DN) hydrogel antifouling coatings.

It is noteworthy that fully physical cross-linked double-network hydrogels demonstrate some potential functions (e.g., self-healing, self-cleaning, and antifreezing) which would help utilization of the hydrogels for long-term antifouling applications in harsh marine environments. For example, Chen et al. [54] designed fully physically cross-linked agar/hydrophobically associated polyacrylamide (HPAM) DN hydrogels, which achieved excellent mechanical strength, high toughness, and notable self-healing property without any external stimuli at room temperature. As illustrated in **Figure 7**, unlike typical DN hydrogels which networks are normally held together with hybrid hydrogen-bond or ionic-covalent interactions, both networks in the designed hydrogels in [54] are physically cross-linked, whereas a hydrogen bond associated agar helix bundles as the first network are interpenetrated with the second network HPAM which are mainly associated by strong hydrophobic interactions between SDS micelles and alkyl groups of stearyl methacrylate (SMA). Fully physical cross-linked DN antifouling hydrogels were further explored by Zheng et al. [59] with a molecular engineering strategy proposed to produce zwitterionic hydrogels with high toughness and self-healing capacity. According to molecular engineering principles, a benzene group and a positively charged imidazole were incorporated into sulfobetaine (SB) monomer to produce a new monomer, 3-(1-(4-vinylbenzyl)-1Himidazol-3-ium-3-yl)propane-1-sulfonate (VBIPS), making the side chain of the VBIPS zwitterionic hydrogels more hydrophobic and strengthening the inter/intra-chain interactions. The formation of dense ionic bonds and π-π interactions between the incorporated benzene and imidazole groups, as well as dynamic nature of these interactions, increase the rigidity of the polymer chain and the interchain interactions, enhancing the tensile and fracture toughness of the hydrogel. The reversible

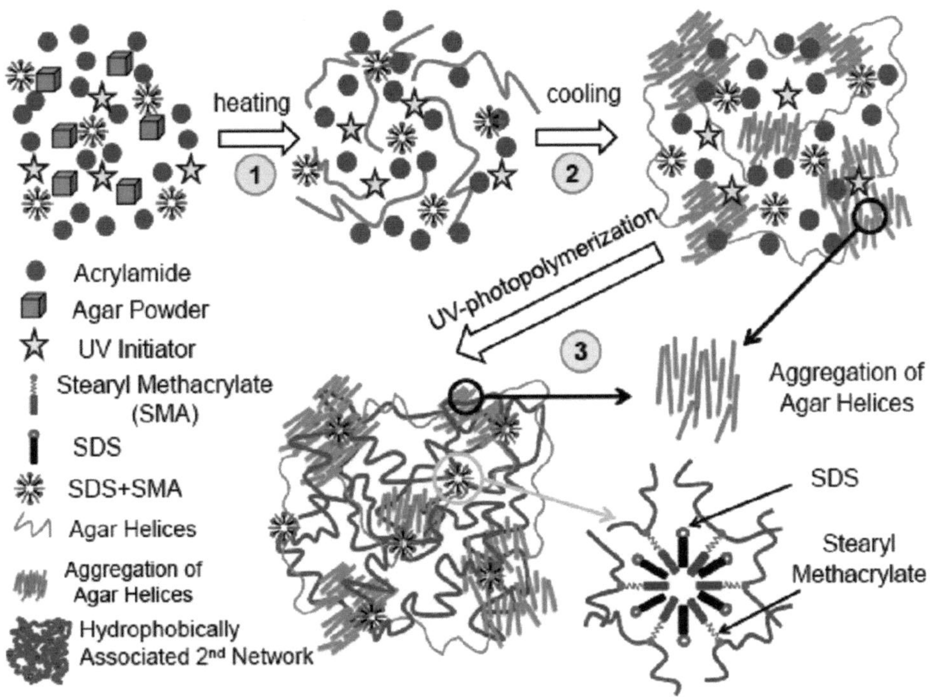

Figure 7.
Illustration of preparation and working principle of fully physically cross-linked agar/HPAM double network hydrogel [54].

and dynamic nature of ionic bonds and π-π interactions also endow the capabilities of self-healing and long-term antifouling via acid or salt treatments.

Furthermore, double network hydrogels could provide a 'repel and kill' effect by incorporating natural antifouling agents as part of networks to achieve long-term antifouling performance, as demonstrated with chitosan in [55, 56] and capsaicin analog in [60].

4.4 Incorporating natural antifoulants and their synthetic analogs into hydrogel coatings

Although the adhesion and mechanical properties of hydrogel coatings can be enhanced via the strategies discussed above, the hydrogel coatings have no intrinsic antibacterial properties and only provide an unsustainable passive biocidal effect through the formation of a hydrophilic hydration layer. This could lead to irreversible marine biofouling and even microbiologically influenced corrosion (MIC) in long-term applications as well as potential risks of marine microorganism transmission. Traditional antifouling coatings containing bactericides or fungicides, such as tributyltin, zinc ions [63], copper oxides [64–66], or TiO_2 [67], exhibit active inhibitory effects on marine organism adhesion but also pose a hazard to marine environment due to their inherent toxicity. Consequently, these coatings are prohibited for use in many countries or deemed unsuitable for practical applications. Incorporating environment-friendly natural antifouling agents or modifying existing natural ones has emerged as a promising method, as aforementioned briefly in Section 4.2. Here, we will further elaborate on recent advances in enhancing the antifouling efficacy of hydrogel coatings while incorporating multifunctional attributes to make them environment-friendly and durable in complex and harsh marine environments [68]. Natural antifoulants can be extracted from marine microorganisms [69], marine plants [70], marine invertebrates, and terrestrial sources [25]. Most natural antifoulant products are biologically extracted organic compounds, intrinsically having better biocompatibility and degradability than heavy metal-based antifouling agents. Ideal antifouling chemicals should possess a range of characteristics, including low dosage yet high efficacy, the ability to inhibit biological adhesion without toxic effects, broad effectiveness against target organisms, and a lack of negative environmental impact during degradation and mass production. Antifoulants usually disrupt their adhesion mechanisms or even kill them when fouling organisms are in close contact with antifoulant-functionalized surfaces [71, 72]. Several antifouling mechanisms have been proposed for natural antifoulants based on their roles, including protein expression regulators, oxidative stress inducers, neurotransmission blockers, surface modifiers, biofilm inhibitors, adhesive production/release inhibitors, and lethal toxicity [25]. Based on their roles and mode of operation, antifouling coatings can be categorized as contact-killing surfaces or antifoulant-releasing surfaces. In the first approach, antimicrobial groups are anchored on the surface, degrading or killing fouling organisms upon contact, while in the second approach, released antifoulants disrupt potential fouling events before attachment occurs [73].

Buzzacchera et al. [74] combined chitosan with PSBMA to develop hydrogel coatings of polymer brushes that exhibited great antifouling capabilities, significantly reducing protein fouling and preventing the activation of platelets and adhesion of white blood cells. The 'repel and kill' effect from incorporating chitosan into double network hydrogel coatings was also demonstrated in **Figure 8** during the antifouling performance assessment in [55, 56].

Chen et al. [75] incorporated 3-(trimethoxysilyl)propyl methacrylate (TMSPMA) into a stable double network PVA/PAM hydrogel for the first time. After hydrolysis, TMSPMA becomes a synthetic orthosilicic acid analog (SOSA) with a structure similar to orthosilicic acid. It is hard for diatoms to distinguish the SOSA from orthosilicic acid, leading them to ingest the SOSA upon attaching to SOSA-contacting hydrogel. However, SOSA cannot be utilized by diatoms to construct cell walls after intake and will disturb their reproduction activity. Therefore, SOSA is often used as an agent against diatom adhesion in the preparation of long-term antifouling hydrogel coatings with an active inhibition effect. He et al. [76] recently introduced a capsaicin analog, N,N′-((4,5,6-trihydroxy-1,3-phenylene) bis(methylene)) dipropenamide (TPA), with other monomers to prepare single network hydrogels. Combined with the antifoulant capsaicin analog TPA with the hydrophilic nature of the as-produced hydrogel, the hydrogel coatings exhibited outstanding algal inhibition and antibacterial performance over 73 days of immersion in a real marine environment. Compared with the double network hydrogel prepared with PVA and another capsaicin analog N-(4-hydroxy-3-methoxybenzyl) acrylamide (HMBA) [60], the SN hydrogel with TPA seems better in the terms of antifouling performance in marine environments, which could be attributed to different antifouling potency of capsaicin analogs used and is worthy of further investigations.

Yang et al. [77] also proposed a facile strategy to construct antifouling hydrogel coatings for long-term biomedical and marine antifouling applications by making use of the synergistic effect of highly hydrated surfaces and the active bactericidal effect from quaternary ammonium cations. The hydrogel coating was prepared with 2-hydroxyethyl methacrylate (HEMA), acrylamide (AM), dimethylaminoethyl acrylate bromoethane (IL-Br), and poly(sodium-p-styrenesulfonate) (PSS). HEMA, AM, and IL-Br copolymerize via terminal double bonds to form hydrophilic polymer chains, which provide a highly hydrophilic surface that greatly imparts antifouling properties to the hydrogel. The quaternary ammonium and sulfonic acid groups from IL-Br and PSS form ionic bonds through electrostatic interactions, while the introduced quaternary ammonium cations enhance the active bactericidal effect. Very recently, Xiong et al. developed an ultrahigh strength SA/PVA/PHMB hydrogel coating [62] and observed that the outstanding long-term antifouling effect in the real

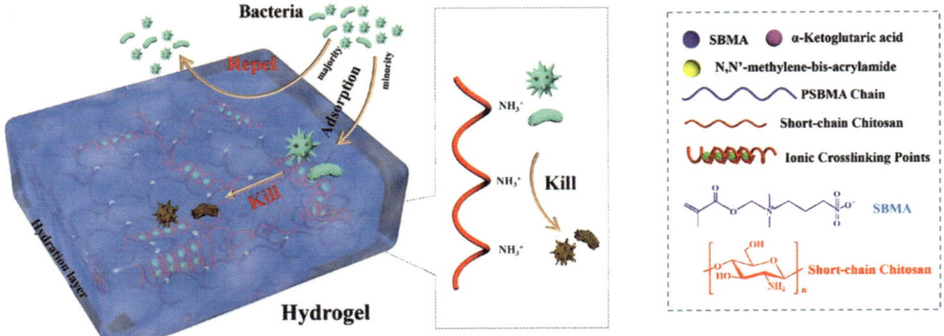

Figure 8.
Illustration of the 'Repel and Kill' effect of hybrid ionic-covalent chitosan/poly(sulfobetaine methacrylate) (CS/PSBMA) DN hydrogel [55].

marine environment was due to strong interactions between the carboxylate anions in SA and the ammonium cations in PHMB as well as the firm bonding of PHMB to the hydrogel. As-designed hydrogel coating combines both excellent mechanical properties and a highly efficient antifouling effect, emerging as one of the most promising marine antifouling coating materials.

4.5 Integration of self-generating function into hydrogel coatings for long-term antifouling performance

Despite good progress made with antifouling coatings, antifouling functions of most reported coatings are readily lost when detachments or scratches occur through physical damage and chemical degradation by water, oxygen, possible catalytic ions, particles, or even rocks and icebergs in harsh marine environments. As proposed in Section 4.1, adhesion to the substrates could be greatly enhanced via surface cross-linking or synergistic multimodal electrostatic interactions. The adhesion enhancement approaches could reduce the coating detachment to some extent but could not fully address detachment and scratch issues caused by physical damages (e.g., erosion by sand particles in seawater or collisions with rocks or marine creatures when naval vessels operate at high speeds). Consequently, it is important to develop durable antifouling coatings with self-polishing, self-generating, self-recovery, or self-healing capabilities. This is especially true when the hydrogel coatings are operated in the mode of contact-killing surfaces which incorporate antifouling or antimicrobial agents that need to be maintained or regenerated on the surfaces. In the case that antifouling agents are released from the coating onto the surface to kill fouling organisms, self-polishing is necessary to maintain an appropriate level of antifouling agents on the surface, as shown in **Figure 9** [78, 79].

At an early stage, most self-polishing coatings contain acrylic or polyurethane copolymers with biocides tributyltin or cuprous oxide (Cu_2O) [80], which degrade slowly in seawater, facilitating the release of fouling biomolecules and organisms. This allows for multiple surface renewals before re-coating is necessary. Such strategies have been extensively used on ship hulls, following significant breakthroughs with the concept of self-peeling hydrogel coating proposed by Xie et al. in [81]. The self-peeling hydrogel was prepared by mixing a polyfunctional aziridine cross-linking agent and a prepolymer copolymerized with methyl methacrylate (MMA), acrylic acid (AA), and tributylsilyl methacrylate (TBSM). The coating can be easily applied onto a surface to form a cross-linked polymer film by conventional brushing or spraying methods. When immersed in seawater, the film self-generates a thin hydrogel layer at the water-contacting surface as a result of the hydrolysis of TBSM. The as-formed hydrogel layer is hydrophilic due to hydration, thus preventing protein adsorption and microorganism attachment. The hydrogel-formation process continues after gradual corrosion and detachment of each top hydrophilic layer, which makes it promising for long-term practical applications. The same group [82, 83] further explored the approach by replacing TBSM with triisopropylsilyl methacrylate (TIPSM) to prepare the self-peelable hydrogel coatings. The field testing results over 2 months of immersion in the sea demonstrated good antibiofouling performance. It was also noted that both the formation and the self-peeling of a thin hydrated layer on the hydrogel coating surface play important roles in its antifouling properties, and there exist delicate balances between the hydrogel formation and the self-peeling rates that can be properly adjusted by the type and content of hydrolysable comonomer and the cross-linking degree. Wu et al. [78] later improved both mechanical

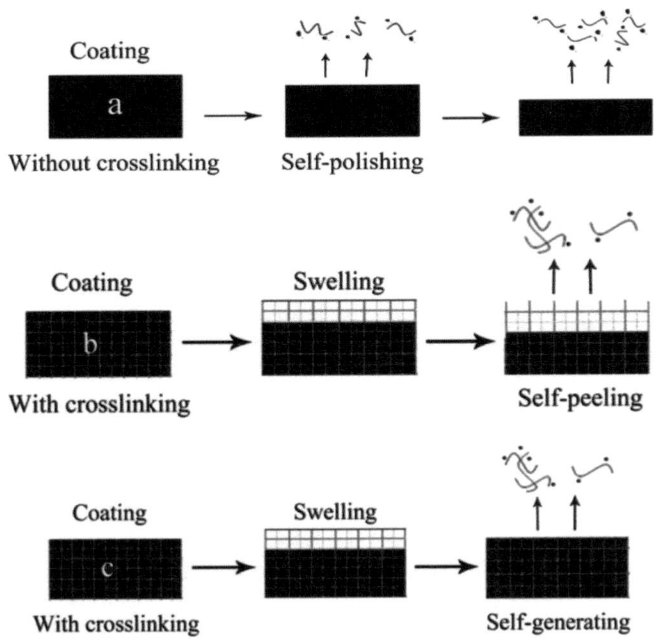

Figure 9.
Schematic of self-polishing processes of (a) uncross-linked and (b) self-peeling hydrogel coatings as well as (c) cross-linked self-generating hydrogel coatings in seawater [78].

and antifouling properties of the coatings by tailoring the used acrylamide (AM) derivatives and incorporating N,N′-[(2-hydroxy-4,5-dimethylbenzene-1,3-diyl)-dimethanediyl] bisprop-2-enamide (HDDE), a capsaicin derivative as active antibacterial agent while maintaining self-peeling and self-generating properties (**Figure 10**).

Dai et al. [84] developed a self-generating and self-renewing zwitterionic hydrogel coating with a different silyl acrylate monomer, tertiary carboxybetaine triisopropylsilyl ethyl acrylate (TCBSA), co-polymerized with MMA and 2-methylene-1,3-dioxepane (MDO). The copolymer rapidly self-generates a zwitterionic surface due to its hydrolysis, and the hydrolyzed polymer chain can dissolve into seawater, leading to a self-renewing dynamic surface, and providing persistent fouling resistance. Moreover, the ester units introduced into the main chain make the polymers degradable, and the degradation rate is sensitive to the enzyme concentration, which further contributes to the self-renewing dynamic surfaces. Lab testing results exhibited excellent protein resistance and antifouling performance against marine bacteria *Pseudomonas* sp. and diatoms. Later the same group [85] developed a degradable hyperbranched polymer containing antifouling moieties N-(2,4,6-trichlorophenyl) maleimide (TCB-TCPM), proposing a coating capable of effectively combating marine biofouling through a synergistic mechanism. Upon exposure to seawater, the polymer surface undergoes hydrolysis, releasing antifouling agents that kill adhered microbes (attack), and generate zwitterionic groups to prevent new organisms from attaching (defense). Moreover, this coating can self-regenerate, with its degradable structure allowing for a continually renewing antifouling surface. The proposed strategy, termed as 'kill–resist–renew trinity', combines these mechanisms to provide a highly promising platform for antifouling coatings. However, its antifouling efficacy needs to be further investigated in actual marine environments for long-term applications. Alternative approaches have also been developed by incorporating

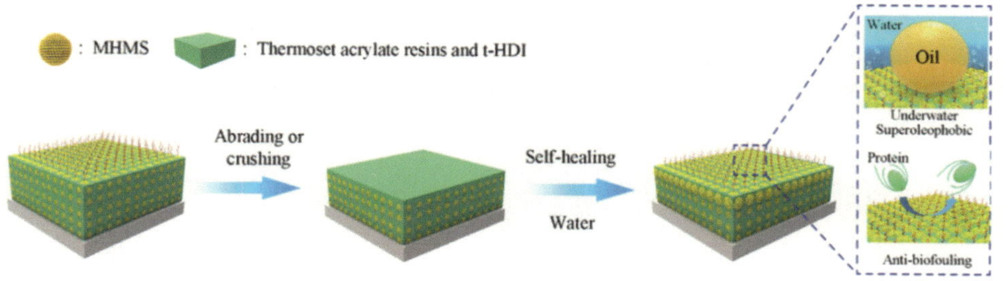

Figure 10.
Working principle of the self-repairing underwater antifouling coating [79].

nanostructured antifouling agents, such as Cu_2O [65, 86], AgNPs [87], and TiO_2 [88], into hydrogel coatings to enable self-cleaning and self-generation for sustainable antifouling in marine environments, although these approaches could raise concerns about potential metallic hazards to marine ecosystem for practical applications.

4.6 Stimuli-responsive hydrogel coatings

Stimuli-responsive coatings can swiftly and reversibly switch their surface properties based on minor environmental changes such as temperature or pH [89]. Combining this responsiveness with antifouling properties could become a valuable strategy for achieving surface regeneration. In such cases, a simple trigger (like salt, pH, temperature, solvent, light, or stress) can easily remove fouling and restore antifouling properties while addressing the concerns caused by incorporating metal oxides antifouling agents for self-regeneration discussed above.

4.6.1 Temperature-responsive antifouling coatings

Thermal-responsive polymers like poly(N-sopropylacrylamide) (PNIPAAm) combine antifouling and fouling release properties. PNIPAAm undergoes a dramatic reversible phase transition from a soluble, swollen, hydrated, and protein-repellent state below its lower critical solution temperature (LCST, 32°C), to an insoluble, collapsed, dehydrated, and protein-adhesive state above this temperature. Below the LCST, antifouling behavior is promoted by the hydration of the acrylamide groups along the polymer brush backbone, which also facilitates the release of dead bacteria. Above the LCST, the layers based on PNIPAAm collapse, causing the coating to absorb smaller proteins and bacteria while exposing antifouling parts that kill bacteria [90]. Yu et al. [91] designed a temperature-responsive switchable surface by grafting the nanopatterned PNIPAM polymer brushes and an antimicrobial quaternary ammonium salt (QAS) at the polymer-free region to achieve the function of attracting bacteria, killing bacterial, followed by releasing bacteria. Wang et al. [92] designed a hierarchical polymer architecture, which can perform the phase transition from antibacterial to antifouling upon increasing temperature from room temperature to physiological conditions at 37°C. The upper layer is a thermal-responsive bactericidal layer of poly(N-isopropylacrylamide-co-2-carboxyethyl acrylate) [P(NIPAAm-co-CEA)], and the lower layer is an antifouling layer of poly(sulfobetaine methacrylate) (PSBMA). At lower temperatures, the upper layer actively kills bacteria, while at higher physiological temperatures, the collapse of the polymer structure exposes the antifouling layer, making it cell-repellent. This dual functionality is particularly useful for medical implants as it prevents infection while minimizing tissue irritation. However, thermal-responsive coatings are seldom reported for marine antifouling applications possibly due to temperature change seldom observed or the impracticability of adjusting environmental temperatures in most real marine applications.

4.6.2 pH-responsive antifouling coatings

pH is a suitable factor to activate a coating to switch between bactericidal and cell-repellency states, thereby preventing the formation of biofilm. The key reason for the antifouling mechanism is associated with pH variation during the metabolization of bacteria [93]. The decrease of pH in the surrounding environment due to bacterial growth acts as a trigger to release bactericidal substances. Xu et al. [94] integrated the pH-responsive strategy, self-cleaning mechanism, and 'one-step' anchoring process to develop environmentally responsive fouling release coatings. The pH-responsive switchable coating was constructed by grafting pH-sensitive poly (2-diisopropyl aminoethyl methacrylate)-b-poly(2-methacryloyloxyethyl phosphorylcholine) (PDPA-b-PMPC) and cationic polylysine (PLYS) chains with azido-modified tannic acid (TA-N3) via copper-free azide–alkyne 'click' reaction. When the pH decreases due to bacterial adhesion onto the surface, the bactericidal polymer brush coating is switched to an antifouling layer with self-defensive (fouling-release/self-cleaning) capability. The coating was found to be stable and durable over 30 days of immersion in filtered seawater or 14 days of exposure to flowing seawater.

4.6.3 Photo-responsive antifouling coatings

While pH and temperature are attractive triggers for some localized marine applications, their uses in naval vessels (e.g., ship hulls) might be limited due to difficulties in temperature and pH control within the exposed local environments. However, due to its noninvasive, inherently clean, and accessible characteristics, light could be an effective stimulus to trigger the coating function to switch reversibly between bactericidal and bacteria-releasing states. Wei et al. [95] explored a smart antibacterial surface that can switch functions in response to UV-visible light. This surface combined azobenzene (Azo) groups with a biocidal β-cyclodextrin derivative conjugated with seven quaternary ammonium salt groups (CD-QAS). When the Azo groups on the coating surface are in trans form, CD-QAS makes the coating surface strongly bactericidal, effectively killing attached bacteria. Upon UV light exposure, Azo groups switch to cis form, and the Azo/CD-QAS inclusion complexes disassociate and release dead bacteria from the surface. This reversible functional switch allows the coating surface to regenerate between killing bacteria and releasing them, which was particularly beneficial for reducing long-term biocidal activity depletion in practical applications. Li et al. [96] developed another antifouling hydrogel coating based on a photothermal antibacterial approach by introducing in-situ modified polydopamine nanoparticles (PDA NPs) with Cu NPs (PDA@Cu NPs) into a polyelectrolyte hydrogel precursor (cationic polyethyleneimine/anionic pectin, CPAP). Due to the inherent bacterial capture/killing capability of the CPAP hydrogel and the enhanced photothermal conversion efficiency of the PDA@Cu NPs, the fabricated coating exhibited highly efficient, convenient, broad-spectrum, and environmentally friendly antibacterial performance. This hydrogel shows excellent potential for antibacterial applications, demonstrating significant efficacy against common pathogens, biocompatibility, and hemocompatibility. However, its application is limited to surfaces easily accessible to light. Moreover, a thick layer of fouling material does not allow light penetration, implying the need for frequent cleaning to maintain photoactive bactericidal capabilities.

4.6.4 Salt-responsive antifouling coatings

Salt-responsive antibacterial hydrogels have been developed for wound-healing applications. For example, Yuan et al. [97] developed a salt-responsive hydrogel with triple functions of antifouling, bactericidal, and bacterial release by combining ε-poly-L-lysine (EPL), poly(ethylene glycol) diglycidyl ether (PEGDGE), and poly(DVBAPS-co-GMA) via a one-pot method. Poly(DVBAPS-co-GMA was firstly copolymerized with monomers (3-(dimethyl(4-vinylbenzyl))ammonium sulfonate (DVBAPS) and glycidyl methacrylate (GMA), which provides antifouling and salt-responsive properties due to the presence of the salt-responsive zwitterionic monomer DVBAPS. Afterward, different functional polymers EPL, poly(DVBAPS-co-GMA), and PEGDGE were mixed with rational feed ratios to react with adjacent chains to form covalently cross-linked hydrogel. Within the hydrogel, EPL as a cationic peptide can impart certain sterilization properties while the PEGDGE with sufficient OEG groups prevents the initial bacterial adhesion effectively. Due to the presence of amine residues, the designed hydrogels were further quaternized by reacting with glycidyltrimethylammonium chloride (GTMAC) to prolong antibacterial properties. The as-prepared hydrogels demonstrated good antifouling and sterilization capabilities, and ca. 94% of the attached bacteria can be released after saline/water switching for several cycles. Fang et al. [98] investigated the antiadhesion properties of salt-sensitive purely zwitterionic hydrogel PSBMA, which was physically self-assembled due to the inter- and intra-molecular ion interactions. The PSBMA polymer was demonstrated to be intriguingly customized into a transient network with outstanding antifouling capability depending on the ion concentration. When ion concentration increases, the PSBMA hydrogel dissociated completely, endowing it as a candidate for bacterial adhesion prevention. Taken these together, salt-responsive hydrogels not only demonstrated their promising potential in wound healing management, but also a strategy to customize for seawater desalination and wastewater treatment applications in marine environments [99].

Stimuli-responsive hydrogel coatings provide valuable guidance to develop long-term and durable antifouling strategies for practical marine applications. However, great challenges remain for the large-scale applicability and durability of stimuli-response hydrogel coatings, as well as the controllability of external stimulus triggers. The multiple switches might also compromise the coating quality, thereby weakening antifouling performance and leading to fouling to build up [100]. Design strategies and the latest research advances in stimuli-responsive antibacterial coatings have been reviewed in more detail [2, 7, 101–103].

5. Conclusions and outlook

The accumulation of marine organisms causes marine fouling, which poses negative impacts on the structural integrity of marine assets and has significant economic and environmental implications. Currently, antifouling coatings represent the most effective strategy to prevent marine fouling. Especially as environmental awareness grows, marine antifouling coatings are evolving toward more effective, durable, and eco-friendly, as well as simpler to process. So far, no single polymer coating formulation has been identified as a universal and effective marine antifouling coating strategy, and leveraging the synergistic advantages of several antifouling strategies into one multifunctional and smarter coating is highly desirable for effectively tackling marine fouling challenges. Among them, hydrogel coatings have been proven promising for practical marine antifouling applications due to their higher efficiency and the diversity of available enhancement strategies, for example, improved adhesion onto any substrates via cross-linking or multimodal electrostatic interactions, toughening via introducing nanomaterials or double network structures, eco-friendliness via integration of natural antifoulant or their analogs, and long-term durability via integration of self-regeneration or self-healing properties.

Despite the great progress made so far, considerable efforts are still needed to further enhance the antifouling efficacy and durability of hydrogel coatings. These efforts should focus on prompting their responsiveness to the changes in the exposure environment and the coating itself (e.g., surface damage/impact, pH, temperature, salinity, and light) as well as developing simplified but precision-controlled mass production technologies, which could be achieved through further development or incorporation of advanced polymerization or processing techniques (e.g., precision in-situ polymerization). For example, transforming these coatings into commercial (large-scale) applications may involve simplifying common covalent grafting methods or shifting the focus to physical adsorption techniques (e.g., layer-by-layer assembly and spray painting). Moreover, integrating smart materials responsive to environmental stimuli (such as changes in pH, temperature, and salinity [104]) could pave the way for dynamic antifouling surfaces that can actively prevent biofouling or release antifouling agents on demand.

Hydrogel coating could effectively prevent marine biofouling and even microbiologically induced corrosion. Nevertheless, the protection of the underlying metallic substrates from marine corrosion is still in demand. As illustrated in **Figure 11(b)**, the hydration layer is hydrophilic, allowing corrosive ionic species such as chloride ions to diffuse through the hydration layer and trigger marine corrosion if the underlying metallic substrates are not adequately protected or if there are cracks or porous defects within hydrogel coating. As such, in order to provide complete protection, antifouling coatings should be integrated with anticorrosion coatings via layer-by-layer structures (e.g., HVOF-deposited amorphous anticorrosion coatings as shown in **Figure 11(a, b)**) or be incorporated with suitable corrosion inhibitors (e.g., BTA inhibitor for Cu-based alloy substrates as shown in **Figure 11(c)**). More efforts should be directed toward further developing integrated antifouling and anticorrosion hydrogel coating systems to achieve comprehensive marine protection, thereby enabling their transformation for practical marine applications.

It is well known that the marine organisms attached to submerged surfaces can not only induce corrosion and cause the marine structural integrity of the affected surfaces compromised but also contribute to the biosecurity risk by spreading invasive species to new regions and disrupting local ecosystems. With further Arctic exploration and the potential increase in Arctic shipping, biofouling on vessels would pose severe threats to the fragile Arctic ecosystem. The sea ice in the Arctic shipping routes could collide with vessel surfaces and damage the surface coatings. Therefore, there are great challenges for marine protection coatings which must have high erosion resistance and antifreezing properties in addition to excellent antifouling and anticorrosion performance under subzero temperatures. Several hydrogel antifouling coatings have been reported with excellent antifreezing properties [59, 106–108], but the mechanical strength of these coating materials is normally in the range of

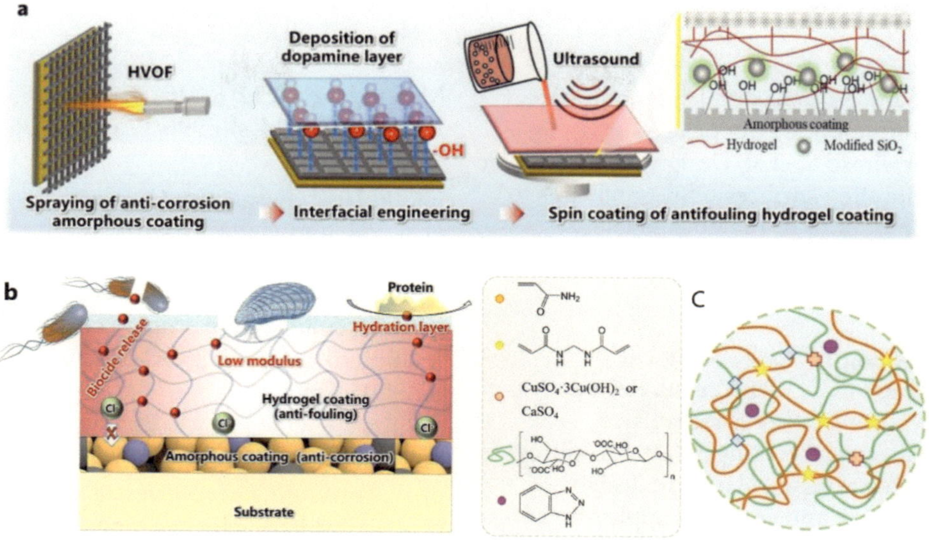

Figure 11.
Schematic illustration of the preparation process of (a) and working principles of integrated antifouling and anticorrosion functionalities of hydrogel-anchored Fe-based amorphous coatings [105]. (c) Illustration of BTA-loaded hydrogel coating [61].

20–190 kPa, which is not strong enough for sustainable and long-term marine antifouling applications regardless of anticorrosion performance. More attention should be paid to developing coatings that can completely protect Arctic shipping facilities from marine fouling and corrosion under subzero temperatures.

Author details

Mengyan Nie* and Koulin Chen
Institute for Materials Discovery, University College London, London, UK

*Address all correspondence to: mengyan.nie@ucl.ac.uk

References

[1] Salta M, Wharton JA, Stoodley P, Dennington SP, Goodes LR, Werwinski S, et al. Designing biomimetic antifouling surfaces. Philosophical Transactions of the Royal Society A. 2010;**368**:4729-4754. DOI: 10.1098/rsta.2010.0195

[2] Zhang T, Liu Q, Meng F, Hou Y, Leung MKH, Wen Y, et al. Recent advances in stimuli-responsive antibacterial coatings: Bacteria-killing and releasing mechanism, design strategies, and potential applications. Progress in Organic Coatings. 2024;**186**:107923. DOI: 10.1016/j.porgcoat.2023.107923

[3] Zhou X, Zheng B. Surface modification for improving immunoassay sensitivity. Lab on a Chip. 2023;**23**:1151-1168. DOI: 10.1039/d2lc00811d

[4] Chen S, Li L, Zhao C, et al. Surface hydration: Principles and applications toward low-fouling/non-fouling biomaterials. Polymer. 2010;**51**:5283-5293. DOI: 10.1016/j.polymer.2010.08.022

[5] Higaki Y, Kobayashi M, Murakami D, et al. Anti-fouling behavior of polymer brush immobilized surfaces. Polymer Journal. 2016;**48**(4):325-331. DOI: 10.1038/pj.2015.137

[6] Lundberg P, Bruin A, Klijnstra JW, et al. Poly (ethylene glycol)-based thiol-ene hydrogel coatings-curing chemistry, aqueous stability, and potential marine antifouling applications. ACS Applied Materials & Interfaces. 2010;**2**(3):903-912. DOI: 10.1021/am900875g

[7] Murosaki T, Ahmed N, Gong JP. Antifouling properties of hydrogels. Science and Technology of Advanced Materials. 2012;**12**(6):064706. DOI: 10.1088/1468-6996/12/6/064706

[8] Ekblad T, Bergström G, Ederth T, et al. Poly (ethylene glycol)-containing hydrogel surfaces for antifouling applications in marine and freshwater environments. Biomacromolecules. 2008;**9**(10):2775-2783. DOI: 10.1021/bm800547m

[9] Murosaki T, Noguchi T, Hashimoto K, et al. Antifouling properties of tough gels against barnacles in a long-term marine environment experiment. Biofouling. 2009;**25**:657-666. DOI: 10.1080/08927010903082628

[10] Murosaki T, Noguchi T, Kakugo A, et al. Antifouling activity of synthetic polymer gels against cyprids of the barnacle (Balanus amphitrite) in vitro. Biofouling. 2009;**25**:313-320. DOI: 10.1080/08927010902730516

[11] Morlat S, Gardette JL. Phototransformation of water-soluble polymers. Part II: Photooxidation of poly(ethylene oxide) in aqueous solution. Polymer. 2003;**44**:7891-7897. DOI: 10.1016/j.polymer.2003.10.022

[12] Han SO, Kim C, Kwon DS. Thermal/oxidative degradation and stabilization of polyethylene glycol. Polymer. 1997;**38**:317-323. DOI: 10.1016/S0032-3861(97)88175-X

[13] Li L, Chen S, Jiang S. Protein interactions with oligo (ethylene glycol) (OEG) self-assembled monolayers: OEG stability, surface packing density and protein adsorption. Journal of Biomaterials Science, Polymer Edition. 2007;**18**(11):1415-1427. DOI: 10.1163/156856207782246795

[14] Leckband D, Sheth S, Halperin A. Grafted poly(ethylene oxide) brushes as nonfouling surface coatings. Journal of Biomaterials Science, Polymer Edition. 1999;**10**(10):1125-1147. DOI: 10.1163/156856299X00720

[15] Wu J, Lin W, Wang Z, Chen S, Chang Y. Investigation of the hydration of nonfouling material poly(sulfobetaine methacrylate) by low-field nuclear magnetic resonance. Langmuir. 2012;**28**(19):7436-7441. DOI: 10.1021/la300394c

[16] Li M, Zhuang B, Yu J. Functional zwitterionic polymers on surface: Structures and applications. Chemistry, an Asian Journal. 2020;**15**:2060-2075. DOI: 10.1002/asia.202000547

[17] Nguyen AT, Baggerman J, Paulusse JM, et al. Stable protein-repellent zwitterionic polymer brushes grafted from silicon nitride. Langmuir. 2011;**27**(6):2587-2594. DOI: 10.1021/la104657c

[18] Schönemann E, Koc J, Aldred N, et al. Synthesis of novel sulfobetaine polymers with differing dipole orientations in their side chains, and their effects on the antifouling properties. Macromolecular Rapid Communications. 2020;**41**(1):1900447. DOI: 10.1002/marc.201900447

[19] Chen S, Zheng J, Li L, et al. Strong resistance of phosphorylcholine self-assembled monolayers to protein adsorption: Insights into nonfouling properties of zwitterionic materials. Journal of the American Chemical Society. 2005;**127**(41):14473-14478. DOI: 10.1021/ja054169u

[20] Chen Z. Surface hydration and antifouling activity of zwitterionic polymers. Langmuir. 2022;**38**(15):4483-4489. DOI: 10.1021/acs.langmuir.2c00512

[21] van Andel E, Lange SC, Pujari SP, et al. Systematic comparison of zwitterionic and non-zwitterionic antifouling polymer brushes on a bead-based platform. Langmuir. 2018;**35**(5):1181-1191. DOI: 10.1021/acs.langmuir.8b01832

[22] Chang Y, Chang W-J, Shih Y-J, et al. Zwitterionic sulfobetaine-grafted poly (vinylidene fluoride) membrane with highly effective blood compatibility via atmospheric plasma-induced surface copolymerization. ACS Applied Materials & Interfaces. 2011;**3**(4):1228-1237. DOI: 10.1021/am200055k

[23] Cai N, Li Q, Zhang J, et al. Antifouling zwitterionic hydrogel coating improves hemocompatibility of activated carbon hemoadsorbent. Journal of Colloid and Interface science. 2017;**503**:168-177

[24] Matsuno R, Ishihara K. Integrated functional nanocolloids covered with artificial cell membranes for biomedical applications. Nano Today. 2011;**6**(1):61-74. DOI: 10.1016%2Fj.nantod.2010.12.009

[25] Jin H, Tian L, Bing W, et al. Bioinspired marine antifouling coatings: Status, prospects, and future. Progress in Materials Science. 2022;**124**:100889. DOI: 10.1016/j.pmatsci.2021.100889

[26] Karthäuser JF, Koc J, Schönemann E, et al. Optimizing fouling resistance of poly (sulfabetaine)s through backbone and charge separation. Advanced Materials Interfaces. 2022;**9**(33):2200677. DOI: 10.1002/admi.202200677

[27] Yeon DK, Ko S, Jeong S, et al. Oxidation-mediated, zwitterionic polydopamine coatings for marine antifouling applications. Langmuir. 2018;**35**(5):1227-1234. DOI: 10.1021/acs.langmuir.8b03454

[28] Liu H, Ma Z, Yang W, et al. Facile preparation of structured zwitterionic polymer substrate via sub-surface initiated atom transfer radical polymerization and its synergistic marine antifouling investigation. European Polymer Journal. 2019;**112**:146-152. DOI: 10.1016/j.eurpolymj.2018.07.025

[29] Berking BB, Poulladofonou G, Karagrigoriou D, et al. Zwitterionic polymeric sulfur ylides with minimal charge separation open a new generation of antifouling and bactericidal materials. Angewandte Chemie International Edition. 2023;**62**(41):e202308971. DOI: 10.1002/anie.202308971

[30] Zhang Y, Liu S, Li T, et al. Cytocompatible and non-fouling zwitterionic hyaluronic acid-based hydrogels using thiol-ene "click" chemistry for cell encapsulation. Carbohydrate Polymers. 2020;**236**:116021. DOI: 10.1016/j.carbpol.2020.116021

[31] Gingell D, Owens N, Hodge P, et al. Adsorption of a novel fluorescent derivative of a poly (ethylene oxide)/poly (butylene oxide) block copolymer on octadecyl glass studied by total internal reflection fluorescence and interferometry. Journal of Biomedical Materials Research. 1994;**28**(4):505-513. DOI: 10.1002/jbm.820280413

[32] Hanein Y, Pan YV, Ratner BD, et al. Micromachining of non-fouling coatings for bio-MEMS applications. Sensors and Actuators B: Chemical. 2001;**81**(1):49-54. DOI: 10.1016/S0925-4005(01)00925-X

[33] Norrman K, Papra A, Kamounah F, et al. Quantification of grafted poly (ethylene glycol)-silanes on silicon by time-of-flight secondary ion mass spectrometry. Journal of Mass Spectrometry. 2002;**37**(7):699-708. DOI: 10.1002/jms.330

[34] Lan S, Veiseh M, Zhang M. Surface modification of silicon and gold-patterned silicon surfaces for improved biocompatibility and cell patterning selectivity. Biosensors and Bioelectronics. 2005;**20**(9):1697-1708. DOI: 10.1016/j.bios.2004.06.025

[35] Shin E, Lim C, Kang UJ, et al. Mussel-inspired copolyether loop with superior antifouling behavior. Macromolecules. 2020;**53**(9):3551-3562. DOI: 10.1021/acs.macromol.0c00481

[36] Kim M, Park J, Lee KM, et al. Peptidomimetic wet-adhesive PEGtides with synergistic and multimodal hydrogen bonding. Journal of the American Chemical Society. 2022;**144**(14):6261-6269. DOI: 10.1021/jacs.1c11737

[37] Wang XT, Deng X, Zhang TD, et al. A versatile hydrophilic and antifouling coating based on dopamine modified four-arm polyethylene glycol by one-step synthesis method. ACS Macro Letters. 2022;**11**(6):805-812. DOI: 10.1021/acsmacrolett.2c00277

[38] Yang J, Xue B, Zhou Y, et al. Spray-painted hydrogel coating for marine antifouling. Advanced Materials Technologies. 2021;**6**(3):2000911. DOI: 10.1002/admt.202000911

[39] Zhu H-W, Zhang J, Su P, et al. Strong adhesion of poly(vinyl alcohol)-glycerol hydrogels onto metal substrates for ma-rine antifouling applications. Soft Matter. 2019;**16**:709-717. DOI: 10.1039/C9SM01413F

[40] Liu Y, Chang Y, Chang YH, Shih YJ. Preparation of amphiphilic polymer-functionalized carbon nanotubes for low-protein-adsorption surfaces and protein-resistant membranes. ACS Applied Materials & Interfaces. 2010;**2**(12):3642-3647. DOI: 10.1021/am100811q

[41] Wang WY, Shi JY, Wang JL, et al. Preparation and characterization of PEG-g-MWCNTs/PSf nano-hybrid membranes with hydrophilicity and antifouling properties. RSC Advances. 2015;**5**(103):84746-84753. DOI: 10.1039/C5RA16077D

[42] Ashraf PM. Nano CuO incorporated polyethylene glycol hydrogel coating over surface modified polyethylene aquaculture cage nets to combat biofouling. Fishery Technology. 2019;**56**(2):115-124

[43] Tian S, Jiang DY, Pu JB, Sun XF, Li ZM, Wu B, et al. A new hybrid silicone-based antifouling coating with nanocomposite hydrogel for durable antifouling properties. Chemical Engineering Journal. 2019;**370**:1-9. DOI: 10.1016/j.cej.2019.03.185

[44] Zhang X, Feng Y, Gao D, et al. Functionalization of cellulosic hydrogels with Cu_2O@ CuO nanospheres: Toward antifouling applications. Carbohydrate Polymers. 2022;**282**:119136. DOI: 10.1016/j.carbpol.2022.119136

[45] Li HL, Zhang LQ, Zhang XH, Zhu GY, Zheng DC, Luo SW, et al. Self-enhanced antibacterial and antifouling behavior of three-dimensional porous Cu_2O nanoparticles functionalized by an organic-inorganic hybrid matrix. ACS Applied Materials & Interfaces. 2023;**15**(32):38808-38820. DOI: 10.1021/acsami.3c06905

[46] Xiong YK, Fang ZQ, Hu DX, Jiang H, Huang L, Mao QT, et al. Nano-CeO_2-loaded polyzwitterionic double-network high-strength hydrogel for highly enhanced synergistic marine antifouling. ACS Applied Materials & Interfaces. 2023;**15**(32):38795-33880. DOI: 10.1021/acsami.3c06741

[47] Hu JP, Zhang DZ, Li WB, Li Y, Shan GR, Zuo M, et al. Construction of a soft antifouling PAA/PSBMA hydrogel coating with high toughness and low swelling through the dynamic coordination bonding provided by $Al(OH)_3$ nanoparticles. ACS Applied Materials & Interfaces. 2024;**16**(5):6433-6446. DOI: 10.1021/acsami.3c17580

[48] Wei Y, Li W, Liu H, et al. In situ preparation of spindle calcium carbonate-chitosan/poly (vinyl alcohol) anti-biofouling hydrogels inspired by shellfish. Journal of Industrial and Engineering Chemistry. 2023;**121**:499-509. DOI: 10.1016/j.jiec.2023.02.005

[49] Zhao W, Zhu Y, Zhang J, et al. A comprehensive study and comparison of four types of zwitterionic hydrogels. Journal of Materials Science. 2018;**53**(19):13813-13825. DOI: 10.1007/s10853-018-2535-6

[50] Shen JJ, Du M, Wu ZL, Song YH, Zheng Q. Strategy to construct polyzwitterionic hydrogel coating with antifouling, drag-reducing and weak swelling performance. RSC Advances. 2019;**9**(4):2081-2091. DOI: 10.1039/C8RA09358J

[51] Gong JP, Katsuyama Y, Kurokawa T, et al. Double-network hydrogels with extremely high mechanical strength. Advanced Materials. 2003;**15**(14):1155-1158. DOI: 10.1002/adma.200304907

[52] Chen H, Chen Q, Hu RD, Wang H, Newby BMZ, Chang Y, et al. Mechanically strong hybrid double network hydrogels with antifouling properties. Journal of Materials Chemistry B. 2015;**3**(27):5426-5435. DOI: 10.1039/c5tb00681c

[53] Jiang DY, Liu ZX, He XY, Han J, Wu XD. Polyacrylamide strengthened mixed-charge hydrogels and their

applications in resistance to protein adsorption and algae attachment. RSC Advances. 2016;**6**(53):47349-47356. DOI: 10.1039/c6ra05312b

[54] Chen Q, Zhu L, Chen H, Yan H, Huang L, Yang J, et al. A novel design strategy for fully physically linked double network hydrogels with tough, fatigue resistant, and self-healing properties. Advanced Functional Materials. 2015;**25**:1598-1607. DOI: 10.1002/adfm.201404357

[55] Zhang J, Shen B, Chen LD, Chen LQ, Mo JY, Feng J. Antibacterial and antifouling hybrid ionic-covalent hydrogels with tunable mechanical properties. ACS Applied Materials & Interfaces. 2019;**11**(35):31594-31604. DOI: 10.1021/acsami.9b08870

[56] Zhang J, Chen LD, Shen B, Chen LQ, Feng J. Ultra-high strength poly(N-(2-hydroxyethyl)acrylamide)/chitosan hydrogel with "repelling and killing" bacteria property. Carbohydrate Polymers. 2019;**225**:115160. DOI: 10.1016/j.carbpol.2019.115160

[57] Zhang J, Peng P, Chen LD, Zhao LP, Feng J. Antifouling poly(N-(2-hydroxyethyl)acrylamide)/sodium alginate double network hydrogels with eminent mechanical properties. Polymer Testing. 2021;**95**:9. DOI: 10.1016/j.polymertesting.2021.107087

[58] Zhang J, Chen L, Chen L, et al. Highly antifouling, biocompatible and tough double network hydrogel based on carboxybetaine-type zwitterionic polymer and alginate. Carbohydrate Polymers. 2021;**257**:117627. DOI: 10.1016/j.carbpol.2021.117627

[59] Zheng SY, Mao SH, Yuan JF, Wang SB, He XM, Zhang XN, et al. Molecularly engineered zwitterionic hydrogels with high toughness and self-healing capacity for soft electronics applications. Chemistry of Materials. 2021;**33**(21):8418-8429. DOI: 10.1021/acs.chemmater.1c02781

[60] Li P, Su X, Hao DZ, Yang M, Gui TJ, Cong WW, et al. One-pot method for preparation of capsaicin-containing double-network hydrogels for marine antifouling. RSC Advances. 2022;**12**(24):15613-15622. DOI: 10.1039/d2ra00502f

[61] Li XLM, Gou JL, Feng HM, Sun X, Qin XD, Wang W, et al. Improved antifouling ability for double-network hydrogel coatings with excellent elastic and toughness under marine tidal environment. Advanced Engineering Materials. 2023;**25**(10):10. DOI: 10.1002/adem.202201801

[62] Xiong YK, Hu DX, Huang L, Fang ZQ, Jiang H, Mao QT, et al. Ultra-high strength sodium alginate/PVA/PHMB double-network hydrogels for marine antifouling. Progress in Organic Coating. 2024;**187**:10. DOI: 10.1016/j.porgcoat.2023.108175

[63] Mu YB, Wu ZL, Ma Y, Zheng JY, Zhang W, Sun ZY, et al. Robust mussel-inspired coatings for controlled zinc ion release. Journal of Materials Chemistry. B. 2017;**5**(9):1742-1752. DOI: 10.1039/c6tb03176e

[64] Ashraf PM, Edwin L. Nano copper oxide incorporated polyethylene glycol hydrogel: An efficient antifouling coating for cage fishing net. International Biodeterioration & Biodegradation. 2016;**115**:39-48. DOI: 10.1016/j.ibiod.2016.07.015

[65] Lin X, Huang X, Zeng C, et al. Poly (vinyl alcohol) hydrogels integrated with cuprous oxide–tannic acid submicroparticles for enhanced mechanical properties and synergetic

antibiofouling. Journal of Colloid and Interface Science. 2019;**535**:491-498. DOI: 10.1016/j.jcis.2018.10.017

[66] Zhang X, Feng YC, Gao DC, Ma W, Jin CZ, Jiang XM, et al. Functionalization of cellulosic hydrogels with Cu_2O@CuO nanospheres: Toward antifouling applications. Carbohydrate Polymers. 2022;**282**:9. DOI: 10.1016/j.carbpol.2022.119136

[67] Natarajan S, Lakshmi DS, Thiagarajan V, et al. Antifouling and anti-algal effects of chitosan nanocomposite (TiO_2/Ag) and pristine (TiO_2 and Ag) films on marine microalgae Dunaliella salina. Journal of Environmental Chemical Engineering. 2018;**6**(6):6870-6880. DOI: 10.1016/j.jece.2018.10.050

[68] Gu Y, Yu L, Mou J, et al. Research strategies to develop environmentally friendly marine antifouling coatings. Marine Drugs. 2020;**18**(7):371. DOI: 10.3390/md18070371

[69] Kyei SK, Darko G, Akaranta O. Chemistry and application of emerging ecofriendly antifouling paints: A review. Journal of Coatings Technology and Research. 2020;**17**:315-332. DOI: 10.1007/s11998-019-00294-3

[70] Pinteus S, Lemos MF, Alves C, et al. The marine invasive seaweeds Asparagopsis armata and Sargassum muticum as targets for greener antifouling solutions. Science of the Total Environment. 2021;**750**:141372. DOI: 10.1016/j.scitotenv.2020.141372

[71] Josphine J, Manjusha W, Subathra C. Evaluation of antifouling potential of staphylococcus sp. isolated from marine sea water. Annals of the Romanian Society for Cell Biology. 2021;**25**(4):6650-6661

[72] Banerjee I, Pangule RC, Kane RS. Antifouling coatings: Recent developments in the design of surfaces that prevent fouling by proteins, bacteria, and marine organisms. Advanced Materials. 2011;**23**(6):690-718. DOI: 10.1002/adma.201001215

[73] Cloutier M, Mantovani D, Rosei F. Antibacterial coatings: Challenges, perspectives, and opportunities. Trends in Biotechnology. 2015;**33**(11):637-652. DOI: 10.1016/j.tibtech.2015.09.002

[74] Buzzacchera I, Vorobii M, Kostina NY, et al. Polymer brush-functionalized chitosan hydrogels as antifouling implant coatings. Biomacromolecules. 2017;**18**(6):1983-1992. DOI: 10.1021/acs.biomac.7b00516

[75] Chen WP, Hao DZ, Guo XL, Hao WJ, Jiang L. Preventing diatom adhesion using a hydrogel with an orthosilicic acid analog as a deceptive food. Journal of Materials Chemistry A. 2018;**6**(39):19125-19132. DOI: 10.1039/c8ta07042c

[76] He G, Liu Y, Liu W, et al. A highly antifouling and eco-friendly hydrogel coating based on capsaicin derivative-functionalized polymer. Journal of Cleaner Production. 2023;**429**:139538. DOI: 10.1016/j.jclepro.2023.139538

[77] Yang Q, Zhou Q, Guo ZY, Song LN, Meng FD, Tong ZM, et al. A facile strategy to construct anti-swelling, antibacterial, and antifogging coatings for protection of medical goggles. Macromolecular Bioscience. 2023;**23**(10):10. DOI: 10.1002/mabi.202300099

[78] Wu G, Li CC, Jiang XH, Yu LM. Highly efficient antifouling property based on self-generating hydrogel layer of polyacrylamide coatings. Journal of Applied Polymer Science. 2016;**133**(42):11. DOI: 10.1002/app.44111

[79] Chen K, Zhou S, Wu L. Self-healing underwater superoleophobic and antibiofouling coatings based on the assembly of hierarchical microgel spheres. ACS Nano. 2016;**10**(1):1386-1394. DOI: 10.1021/acsnano.5b06816

[80] Kiil S, Dam-Johansen K, Weinell CE, et al. Dynamic simulations of a self-polishing antifouling paint exposed to seawater. Journal of Coatings Technology. 2002;**74**(929):45-54. DOI: 10.1007/BF02698368

[81] Xie LY, Hong F, He CX, Ma CF, Liu JH, Zhang GZ, et al. Coatings with a self-generating hydrogel surface for antifouling. Polymer. 2011;**52**(17):3738-3744. DOI: 10.1016/j.polymer.2011.06.033

[82] Hong F, Xie LY, He CX, Liu JH, Zhang GZ, Wu C. Effects of hydrolyzable comonomer and cross-linking on anti-biofouling terpolymer coatings. Polymer. 2013;**54**(12):2966-2972. DOI: 10.1016/j.polymer.2013.04.010

[83] Hong F, Xie LY, He CX, Liu JH, Zhang GZ, Wu C. Novel hybrid anti-biofouling coatings with a self-peeling and self-generated micro-structured soft and dynamic surface. Journal of Materials Chemistry. B. 2013;**1**(15):2048-2055. DOI: 10.1039/c3tb00031a

[84] Dai G, Xie Q, Ai X, et al. Self-generating and self-renewing zwitterionic polymer surfaces for marine anti-biofouling. ACS Applied Materials & Interfaces. 2019;**11**(44):41750-41757. DOI: 10.1021/acsami.9b16775

[85] Dai G, Ai X, Mei L, et al. Kill–resist–renew trinity: Hyperbranched polymer with self-regenerating attack and defense for antifouling coatings. ACS Applied Materials & Interfaces. 2021;**13**(11):13735-13743. DOI: 10.1021/acsami.1c02273

[86] Wang YD, Ren XT, Ma XB, Xue LL, Ding FY. A durable and self-cleaning hydrogel micro-powder modified coating with improved utilization of Cu^{2+} for marine antifouling. Journal of Polymer Research. 2022;**29**(6):11. DOI: 10.1007/s10965-022-02940-8

[87] Liu YT, Zhao SS, Jiang YJ, Liu Q, Liu Z, Song CX, et al. AgNP hybrid organosilicon antifouling coating with spontaneous self-healing and long-term antifouling properties. Industrial and Engineering Chemistry Research. 2023;**62**(33):12999-13008. DOI: 10.1021/acs.iecr.3c01569

[88] Xu XF, Zhu TX, Zheng WW, Xian CY, Huang JY, Chen Z, et al. A robust and transparent hydrogel coating for sustainable antifogging with excellent self-cleaning and self-healing ability. Chemical Engineering Journal. 2023;**451**:9. DOI: 10.1016/j.cej.2022.137879

[89] Burkert S, Bittrich E, Kuntzsch M, et al. Protein resistance of PNIPAAm brushes: Application to switchable protein adsorption. Langmuir. 2010;**26**(3):1786-1795. DOI: 10.1021/la902505q

[90] Huber DL, Manginell RP, Samara MA, et al. Programmed adsorption and release of proteins in a microfluidic device. Science. 2003;**301**(5631):352-354. DOI: 10.1126/science.1080759

[91] Yu Q, Cho J, Shivapooja P, et al. Nanopatterned smart polymer surfaces for controlled attachment, killing, and release of bacteria. ACS Applied Materials & Interfaces. 2013;**5**(19):9295-9304. DOI: 10.1021/am4022279

[92] Wang X, Yan S, Song L, et al. Temperature-responsive hierarchical polymer brushes switching from

bactericidal to cell repellency. ACS Applied Materials & Interfaces. 2017;**9**(46):40930-40939. DOI: 10.1021/acsami.7b09968

[93] Zhuk I, Jariwala F, Attygalle AB, Wu Y, Libera MR, Sukhishvili SA. Self-defensive layer-by-layer films with bacteria-triggered antibiotic release. ACS Nano. 2014;**26**(8):7733-7745. DOI: 10.1021/nn500674g

[94] Xu G, Neoh KG, Kang ET, Teo SLM. Switchable antimicrobial and antifouling coatings from tannic acid-scaffolded binary polymer brushes. ACS Sustainable Chemistry & Engineering. 2020;**8**(6):2586-2595. DOI: 10.1021/acssuschemeng.9b07836

[95] Wei T, Zhan W, Yu Q, et al. Smart biointerface with photoswitched functions between bactericidal activity and bacteria-releasing ability. ACS Applied Materials & Interfaces. 2017;**9**(31):25767-25774. DOI: 10.1021/acsami.7b06483

[96] Li Z, You S, Mao R, et al. Architecting polyelectrolyte hydrogels with Cu-assisted polydopamine nanoparticles for photothermal antibacterial therapy. Materials Today. Bio. 2022;**15**:100264. DOI: 10.1016/j.mtbio.2022.100264

[97] Yuan JF, Zhang D, He XM, Ni YF, Che LB, Wu J, et al. Cationic peptide-based salt-responsive antibacterial hydrogel dressings for wound healing. International Journal of Biological Macromolecules. 2021;**190**:754-762. DOI: 10.1016/j.ijbiomac.2021.09.019

[98] Fang YL, Huang SS, Gong X, King JA, Wang YQ, Zhang JC, et al. Salt sensitive purely zwitterionic physical hydrogel for prevention of postoperative tissue adhesion. Acta Biomaterialia. 2023;**158**:239-251. DOI: 10.1016/j.actbio.2022.12.045

[99] Bhoj Y, Tharmavaram M, Rawtani D. A comprehensive approach to antifouling strategies in desalination, marine environment, and wastewater treatment. Chemical Physics Impact. 2021;**2**:100008. DOI: 10.1016/j.chphi.2020.100008

[100] Maan AMC, Hofman AH, de Vos WM, Kamperman M. Recent developments and practical feasibility of polymer-based antifouling coatings. Advanced Functional Materials. 2020;**30**(32):2000936. DOI: 10.1002/adfm.202000936

[101] Han X, Wu JH, Zhang XH, Shi JY, Wei JX, Yang Y, et al. The progress on antifouling organic coating: From biocide to biomimetic surface. Journal of Materials Science and Technology. 2021;**61**:46-62. DOI: 10.1016/j.jmst.2020.07.002

[102] Su D, Bai XQ, He XY. Research progress on hydrogel materials and their antifouling properties. European Polymer Journal. 2022;**181**:16. DOI: 10.1016/j.eurpolymj.2022.111665

[103] Kio M, Klauda J. Advances in emerging hydrogel fouling-release coatings for marine applications. Journal of Coating Technology and Research. 2024;**21**(3):827-856. DOI: 10.1007/s11998-023-00895-z

[104] Zhang YQ, Wei HL, Hua BY, Hu CW, Zhang WJ. Preparation and application of the thermo-/pH−/ion-sensitive semi-IPN hydrogel based on chitosan. International Journal of Biological Macromolecules. 2024;**258**:14. DOI: 10.1016/j.ijbiomac.2023.128968

[105] Zhang LY, Feng DQ, Zhu PY, Song WL, Yasir M, Zhang C, et al. Hydrogel-anchored Fe-based amorphous coatings with integrated antifouling and anticorrosion functionality. ACS Applied Materials & Interfaces.

2023;**15**(10):13644-13655. DOI: 10.1021/acsami.3c00227

[106] Gao H, Zhao Z, Cai Y, et al. Adaptive and freeze-tolerant heteronetwork organohydrogels with enhanced mechanical stability over a wide temperature range. Nature Communications. 2017;**8**:15911. DOI: 10.1038/ncomms15911

[107] Ma MQ, Zhang C, Chen TT, Yang J, Wang JJ, Ji J, et al. Bioinspired polydopamine/polyzwitterion coatings for underwater anti-oil and -freezing surfaces. Langmuir. 2019;**35**(5):1895-1901. DOI: 10.1021/acs.langmuir.8b02320

[108] Tian SH, Wang MM, Wang X, Wang LY, Yang DZ, Nie J, et al. Smart hydrogel sensors with antifreezing, antifouling properties for wound healing. ACS Biomaterials Science & Engineering. 2022;**8**(5):1867-1877. DOI: 10.1021/acsbiomaterials.1c01599